Discovering Complexity

Discovering Complexity

DECOMPOSITION AND LOCALIZATION AS STRATEGIES IN SCIENTIFIC RESEARCH

William Bechtel
  and
Robert C. Richardson

PRINCETON UNIVERSITY PRESS

PRINCETON, NEW JERSEY

Copyright © 1993 by Princeton University Press
Published by Princeton University Press, 41 William Street,
Princeton, New Jersey 08540
In the United Kingdom: Princeton University Press, Oxford

Library of Congress Cataloging-in-Publication Data

Bechtel, William.
Discovering complexity: decomposition and localization as
strategies in scientific research / William Bechtel, Robert C.
Richardson.
p. cm.
Includes bibliographical references and index.
1. Research—Methodology. 2. Decomposition method.
3. Localization theory. 4. Complexity (Philosophy) I. Richardson,
Robert C., 1949– . II. Title.
Q180.55.M4B4 1992 507′.2—dc20 92-3359

ISBN 0-691-08762-8

This book has been composed in Linotron Caledonia

Princeton University Press books are printed on
acid-free paper, and meet the guidelines for permanence
and durability of the Committee on Production
Guidelines for Book Longevity of the
Council on Library Resources

Printed in the United States of America

10 9 8 7 6 5 4 3 2 1

For Adele

and for Genevieve, Dan, and Annabelle

Contents

Preface   xi

PART I: *Scientific Discovery and Rationality*

CHAPTER 1
Cognitive Strategies and Scientific Discovery   3

   1. Rationalizing Scientific Discovery   3
   2. Procedural Rationality   11

CHAPTER 2
Complex Systems and Mechanistic Explanations   17

   1. Mechanistic Explanation   17
   2. Decomposition and Localization   23
   3. Hierarchy and Organization   27
   4. Conclusion: Failure of Localization   31

PART II: *Emerging Mechanisms*

   Introduction   35

CHAPTER 3
Identifying the Locus of Control   39

   1. Introduction: Identifying System and Context   39
   2. External Control: The Environment as a Control   41
   3. Internal Control: The System as a Control   47
   4. Fixing on a Locus of Control: The Cell in Respiration 51
   5. Conclusion: Localization of Function   59

CHAPTER 4
Direct Localization   63

   1. Introduction: Relocating Control   63
   2. Phrenology and Cerebral Localization   65
   3. Competing Models of Cellular Respiration   72
   4. Conclusion: Direct Localization and Competing Mechanisms 88

CHAPTER 5
The Rejection of Mechanism   93

   1. Introduction: Mechanism and Its Opponents   93
   2. Flourens and the Integrity of the Nervous System   95

3. *The Vitalist Opposition to Mechanistic Physiology*   99
4. *Conclusion: Settling for Descriptions*   113

PART III: *Elaborating Mechanisms*

Introduction   119

CHAPTER 6
Complex Localization   125

1. *Introduction: Constraints on Localization*   125
2. *Top-Down Constraints*   128
3. *Bottom-Up Constraints*   138
4. *Conclusion: The Rise and Decline of Decomposability*   145

CHAPTER 7
Integrated Mechanisms   149

1. *Introduction: Replacing a Direct Localization*   149
2. *Direct Localization of Fermentation in Zymase*   153
3. *A Complex Linear Model of Fermentation*   156
4. *An Integrated System Responsible for Fermentation*   163
5. *Conclusion: The Discovery of Integration*   168

CHAPTER 8
Reconstituting the Phenomena   173

1. *Introduction: Biochemical Genetics*   173
2. *Classical Genetics*   175
3. *Developmental Genetics*   181
4. *One Gene/One Enzyme*   188
5. *Conclusion: Reconstituting the Phenomena*   192

PART IV: *Emergent Mechanism*

Introduction   199

CHAPTER 9
"Emergent" Phenomena in Interconnected Networks   202

1. *Introduction: Dispensing with Modules*   202
2. *Hierarchical Control: Hughlings Jackson's Analysis of the Nervous System*   203
3. *Parallel Distributed Processing and Cognition*   210
4. *Distributed Mechanisms for Genomic Regulation*   223
5. *Conclusion: Mechanistic Explanations without Functional Decomposition and Localization*   227

CHAPTER 10
Constructing Causal Explanations   230
  1. Decomposition and Localization in Perspective   230
  2. Four Constraints on Development   234
  3. Conclusion: Looking Forward   243

Notes   245

References   257

Index   281

Preface

DISCOVERING COMPLEXITY represents the culmination of nearly a decade of collaboration. Our work developed out of struggles with the standard philosophical framework of theory reduction, one which we found nearly impossible to apply to the life sciences. We discovered, instead, that the attempts to localize the causes of phenomena in components of complex systems and to investigate the organization in these systems reflect what many scientists understand by the term *reduction*—a discovery that made more sense of the contrasting approaches of scientists toward their own disciplines.

William Wimsatt, our mentor in graduate school and in the development of our careers, introduced us to the notion of localization and to the various and complex problems affecting reduction. Under his instruction we became familiar with Elliot Valentine's text *Brain Control: A Critical Examination of Brain Stimulation and Psychosurgery* (New York: Wiley, 1973) and with the work of R. L. Gregory, whose research on localization in the neurosciences figures prominently in Wimsatt's work as well as in our own. Finally, Wimsatt led us to the work of Herbert Simon, whose ideas constitute a dominant theme in our book.

We began working on this book, originally entitled "An Alternative Approach to Reductive Explanation in the Philosophy of Science: A Study of Localizationist Research Programs," in 1983. The completed project took far longer than either of us anticipated. This was due in part to the extensive digesting required by the historical investigations; our initial analyses of the historical cases were far too detailed for the level of study we could pursue. Moreover, we had originally anticipated simply presenting parallel historical narratives from the neurosciences, physiological chemistry, and genetics, but subsequently decided that it was more fruitful and interesting to organize the presentation conceptually, incorporating the appropriate pieces of each study into such a framework.

The extra time has been used in part to deepen our knowledge of the history and to reconceptualize the problem. While we started with a focus on reduction, the notion (as Marjorie Grene will be pleased to note) has all but disappeared from the book. Our focus shifted to the process of theory development and change. As we pursued the historical cases, we found that traditional philosophical accounts of theories did not fit. The scientists we were studying were not primarily interested in general laws; they sought instead to identify causal components that explained how various systems produced specific phenomena. For this mode of explanation we introduced the notion of a *mechanistic explanation*. Given that we could not appeal in our treatment to more familiar philosophical notions, we had to develop our own conceptual framework. For those who are interested in the differences between our approach and more traditional philosophical analyses, we draw some contrasts in Chapter 10.

The structure of *Discovering Complexity* requires some comment. The case studies found in chapters 3 through 9 are the core of the book. It was through the analyses of these cases that our understanding of mechanistic explanation, and the process of developing such explanations, took form, and it is on them that we rest our claim to have provided a realistic analysis of scientific discovery. However, many readers may prefer to read only the conceptual framework at which we have arrived, without the historical detail; those so inclined should concentrate on chapters 1, 2, and 10. We believe that the details matter, and it is on the details rather than the general picture that we rest our case. For those with little patience for details, this will be a much shorter book. Chapter 1 argues for the focus on the dynamics of scientific discovery and offers our case for psychologizing the discovery process. Chapter 2 introduces the heuristics of decomposition and localization—which we claim guide much scientific theorizing directed toward mechanistic explanations—and explores the assumptions these heuristics entail. Chapter 10 then places the psychologistic analysis into a somewhat broader framework by introducing additional constraints that must be analyzed in a full dynamic analysis of the development of mechanistic explanations.

The case studies are drawn largely from the nineteenth- and twentieth-century history of disciplines focused on the operation of various components of living organisms. We concentrate primarily on episodes in the development of neurophysiological explanations of mental activities and of chemical accounts of basic physiological processes such as fermentation and oxidation. While we have omitted much historical detail, these cases are admittedly still challenging to a reader not familiar with the history of the scientific disciplines involved. To somewhat mitigate the difficulty for those who do not know the history, we have included introductions to Parts II, III, and IV that both provide overviews of the cases and make clear what conceptual points are demonstrated in each. A central piece of our philosophical framework is the identification of a number of *choice-points* at which researchers make decisions that affect the nature of their research programs. We discuss such points throughout Parts II and III and represent them in a flowchart that is further elaborated at the end of Chapters 3–8.

We take the history of science seriously. We are, however, philosophers—not historians—by training. While we have consulted many of the original publications in which scientists developed theories and reported their experiments, in most cases we have not gone behind the scenes ourselves to do original historical research. Rather, we have relied on, and learned from, numerous historians whose work we identify in the text. We respect their skills at historiography and view our contribution to advancing a model of the development of mechanistic explanations as complementary to their work. Moreover, although focusing on psychological factors guiding theory development, we make no claim that these are the only important factors. Indeed, we hold the opposite view: a variety of factors are at work, many of them social in nature. We have simply chosen one set of constraints on theory development and have tried to explore its operation. A complete analysis of the dynamics of theory development, however, must integrate this aspect with many other perspectives; we sketch, in Chapter 10, how some of these might fit into a more complete dynamic analysis.

Many people and institutions have supported us in this endeavor. We have benefited both from William Wimsatt's friendship and from our many discussions of conceptual issues and historical cases. We were also inspired by Marjorie Grene and Richard Burian when we attended their institute on philosophy of biology, which was supported by the National Endowment for the Humanities, at Cornell University in the summer of 1982. Marjorie goaded, chided, and sometimes scolded us over the subsequent years. We both hold her in highest esteem. Dick Burian has been a constant reminder of the breadth and depth that is necessary to do serious philosophy of science. Lindley Darden has read numerous drafts of this book and struggled with more biochemistry and neuroscience than she ever cared to. She has continuously prodded us to make the main messages more explicit, and we hope she will find her efforts rewarded. Jane Maienshein has not only provided detailed comments on the entire manuscript, but has discussed the issues of this project with us over many years. To these five scholars we owe our deepest gratitude, not only for their help but for their friendship.

We have benefited from discussions on topics relevant to this project with many other colleagues. Their support and contributions are hard to overestimate. These people include especially Adele Abrahamsen, Robert Brandon, Steve Fuller, Rebecca German, Donald Gustafson, Thomas Kane, Stuart Kauffman, William G. Lycan, Robert McCauley, William E. Morris, Michael Ruse, Miriam Solomon, and Paul Thagard. The illustrations for the text were prepared by Linda Weiner Morris, for whose patience and skill we are grateful.

We have also benefited from the help and support of numerous institutions. Grant RO-20758 from the National Endowment for the Humanities, which we received when we began this project in 1983, provided the resources for the historical investigations—as well as the time for reflection, research, and collaboration—that led to the initial drafting of this project. We are both grateful for support from the NEH and the National Science Foundation for other projects that have left their mark on this book. Both of our institutions have also been generous in support. In particular, we would like to thank the Taft Faculty Committee at the University of Cincinnati and the Research Grant Programs at Georgia State. And, finally, we would like to mention the summer courses taken at the Marine Biological Laboratory in Woods Hole, which helped enormously with our research on the history of biology.

PART I

Scientific Discovery and Rationality

The issue is not theory-using, but theory-finding; my concern is
not with the testing of hypotheses, but with their discovery.
Let us examine not how observation, facts and data are built up
into general systems of physical explanation, but how these
systems are built into our observations, and our
appreciation of facts and data.

—Hanson 1958

Scientific discovery is a form of problem solving, and . . . the
processes whereby science is carried on can be explained in
the terms that have been used to explain the processes
of problem solving.

—Simon 1966

CHAPTER 1

Cognitive Strategies and Scientific Discovery

> We have come, or are coming, at last to the end of this epoch, the
> epoch presided over by the concepts of Newtonian cosmology and
> Newtonian method. We are in the midst of a new philosophical
> revolution, a revolution in which, indeed, the new physics too has
> had due influence, but a revolution founded squarely on the
> disciplines concerned with life: on biology, psychology,
> sociology, history, even theology and art criticism.
> —M. Grene 1974

1. RATIONALIZING SCIENTIFIC DISCOVERY

Logical positivism drew a principled and sharp distinction between the
context of justification and that of discovery. According to this view, the
empirical evaluation of scientific theories could be submitted to logical
analysis, with the goal of specifying the conditions under which a theory
would be confirmed or disconfirmed. Theories were accordingly given an
axiomatic rendering. The fundamental laws provided the axioms in terms
of which all else was explained. The theorems were either nonfundamen-
tal laws or observational claims. Scientific discovery, by way of contrast,
seemed to be far less congenial to logical analysis and was therefore
shunted off as a "merely" psychological matter having no significant consequences for a theory of scientific rationality. Hans Reichenbach captured
the spirit of the times in his classic work, *The Rise of Scientific Philosophy*:

> The act of discovery escapes logical analysis; there are no logical rules in terms
> of which a 'discovery machine' could be constructed that would take over the
> creative function of the genius. But it is not the logician's task to account for
> scientific discoveries; all he can do is to analyze the relation between given facts
> and a theory presented to him with the claim that it explains these facts. In
> other words, logic is concerned with the context of justification. (1951, p. 231)

Discovery correspondingly was seen as a mystical process; justification
alone was susceptible to logical analysis. Even those like Karl Popper,
who dogmatically rejected the idea that scientific theories could ever be
verified, held on to the distinction between what could be logically analyzed and what had to be left to psychology. In the opening chapter of *The
Logic of Scientific Discovery*, Popper writes:

> The initial stage, the act of conceiving or inventing a theory, seems to me nei-
> ther to call for logical analysis nor to be susceptible of it. The question how it
> happens that a new idea occurs to a man . . . may be of great interest to empir-
> ical psychology; but it is irrelevant to the logical analysis of scientific knowledge.
> This latter is concerned not with *questions of fact*, . . . but only with questions
> of *justification or validity*. (1959, p. 31)

Scientific discovery is, ironically, left aside as a topic for psychology, but
not for logic.[1]

More recent philosophy of science, together with the branches of cognitive science, has returned to the investigation of discovery, recognizing
that the distinction between discovery and justification is artificial and
that there are rational or 'logical' considerations guiding both processes
(for example, see Darden 1980 and 1991; Giere 1979, 1988; Nickles 1980,
1982, 1987a, 1987b; Simon 1977; and Thagard 1988 and 1992). Pressure
against any firm distinction between justification and discovery correspondingly comes from two directions. One challenges the purity of a
'logic' of justification; treating justification in abstraction from psychological processes, social context, and historical setting also abstracts from the
scientific practice that gives it life. The other demystifies the process of
discovery; once recognized for the problem-solving process it is, discovery too is animated by scientific practice. If discovery and justification are
not quite one, they are at least reflections of a single activity that finds its
expression in scientific practice—though not, of course, *only* in scientific
practice.

Positivistic accounts not only ignore discovery, but also see justification
in exclusively empirical terms. Theory confirmation and disconfirmation
involve more than simple considerations of empirical adequacy—empirical data has unclear relevance in the absence of an appropriate theory.[2]
Scientists were uncertain of the significance of, for example, the data on
the relative frequencies of neucleotide bases prior to the recognition of
complementary binding in Watson's and Crick's double helix. And Mendel's hybridization ratios did not have the significance they have for us
before the study of genetics was divorced from embryology and development in the early decades of the twentieth century. Empirical data, moreover, may be regarded as irrelevant even when it cannot be reconciled
with a favored theory. It is well known, for example, that the deviations in
the orbit of Mercury were observed and recognized, but largely ignored,
before the advent of the theory of relativity; similarly, despite considerable evidence for maternal inheritance, cytoplasmic inheritance found no
significant place within a theory committed to nuclear inheritance, and
evidence of maternal inheritance was dismissed as irrelevant rather than
disconfirmatory.[3] Empirical data also may be reconciled with a theory by

a variety of apparently ad hoc devices, as discontinuities in the fossil record were dismissed in nineteenth-century evolutionary debates. Empirical constraints are simply not enough to explain theoretical relevance. If we are to understand confirmation or disconfirmation as it occurs in scientific practice, we must seek additional constraints from other sources. Likewise, while empirical constraints are important to discovery, they are not sufficient here either. A model capable of accommodating either confirmation or discovery must incorporate empirical constraints, but it must incorporate further constraints as well.

There is a wide range of constraints that are potentially relevant, including some that are broadly social, historical, or technological. That psychological constraints are also at least *part* of what is needed for a theory of justification is suggested by the observation that logicist accounts of justification operate with unrealistic assumptions concerning what human beings can do. For example, despite considerable Bayesian literature on confirmation, it is reasonably clear that humans systematically undervalue prior probability, to the point of ignoring it altogether, even when given "data" that is entirely worthless. In psychological experiments, humans violate fundamental rules of probability, ignore base rates, and neglect questions of sample size in making predictions (Kahneman and Tversky 1972, 1973; Tversky and Kahneman 1974), fail to seek disconfirmatory evidence even when it is readily obtained, do not use available disconfirming evidence, and ignore alternative explanations of the available data (Johnson-Laird and Wason 1970; Wason 1966). Though one might wish it were not so, scientists do little better (see Faust 1984). Clinical prediction has long been known to be less reliable than actuarial judgment: in a variety of areas, from neuropsychology to psychiatry, it is clear that we can get more reliable prediction and diagnosis by using mechanical means than by relying on human judgment. Nonetheless, there is a pervasive tendency to overestimate human judgment's usefulness. For example, clinicians given results generated by actuarial means will place undue reliance on their own intuitive judgment, discounting the statistical data. Such discretionary judgment reduces reliability.[4] Moreover, as Amos Tversky and Daniel Kahneman (1971) argue, the situation is not appreciably improved by turning to individuals trained in statistics (for some mitigating considerations, see Nisbett 1988). Even statistically sophisticated psychologists adopt sample sizes in replication studies that are unreasonably low, and *then* underestimate the significance of the results. If formalist theories prescribe procedures that are fundamentally different from those constitutive of human reasoning, it should be little surprise when their prescriptions diverge from what we actually do. Yet science is a human enterprise; if the problem is understanding the processes of human confirmation and human discovery, we must evidently look to human reasoning.

We propose to treat scientific change, scientific discovery, and scientific rationality in a way that is simultaneously developmental and psychological. Our approach is *developmental* in emphasizing the dynamics of theoretical change rather than its statics and in embracing the historical contingency of the resulting theories. In positivist and neopositivist models of theories, questions of the dynamics of theories are treated as derivative from their static descriptions. Theories are taken as axiomatized systems. Convergence and commensurability are the dynamic variables, and they, in turn, reduce to similarity in theoretical commitments. The neopositivist emphasis is on theories abstracted from their development. By contrast, we seek the laws of motion. It is in modification and change, in theories in disequilibrium and flux, that we can hope to reveal the cognitive strategies that drive scientific change and are constitutive of scientific rationality. Classical models of theories were self-consciously ahistorical, divorcing contemporary content from its origins. Our emphasis on dynamics leaves no room for such luxury. To shift the image from the physical sciences to the biological, we look to the ontogeny of theories. In the ontogeny of an organism, an earlier developmental stage may exemplify structures that have no counterparts in the adult organism. Nevertheless, these stages are crucial to development, and the resulting structures cannot be understood in abstraction from their sources in earlier developmental stages: the obsolete earlier stages make the later stages possible. In parallel fashion, later theoretical developments in a scientific discipline may have little in common with their precursors. Nevertheless, these early stages may be crucial to the emerging discipline, and it is not generally possible to understand the theoretical commitments at a given time in abstraction from the commitments of precursors. In the ontogeny of theories, no less than in the ontogeny of organisms, historical contingency is the consequence of developmental priority (cf. Wimsatt 1986b).

Our approach is *psychological* in emphasizing that theoretical development is in part an expression of human cognitive style, a consequence of the typical strategies with which human beings attack problems, and the cognitive limitations that make these strategies necessary. Discovery and justification alike should be understood in psychological terms. A model of scientific discovery should treat scientific problem-solving as a special case of human problem-solving. As such, it is subject to all the contingencies of human problem-solving, both internal and external, psychological and social. The approach to theoretical, explanatory, and empirical problems posed by scientific investigation is determined, in part at least, by cognitive capacities and limitations. Humans do *not* perform exhaustive searches, even when the problems faced are simple enough to make it feasible. Humans do *not* operate with axiomatized structures, searching the consequence set for confirmation and disconfirmation. Explanatory models, simplified schemata, and the assimilation of complex phenomena to these prototypes is typical of human cognitive style (cf. Cartwright 1983; Johnson-Laird 1983; Laymon 1980, 1982, 1989; McCauley 1985; Richardson 1986b, c). The process is one of adjustment and pattern matching, rather than analysis and deduction. Search is selective and limited, guided by heuristics, local in its application, and biased in its outcome. To psychologize the treatment of scientific change is to take these issues of cognitive style seriously. The theorist is, after all, important to the theory.

Our enterprise is, thus, not only psychological, but self-consciously *psychologistic*. This does not mean it is wholly descriptive. We operate on the assumptions that psychological constraints are important in understanding scientific change, confirmation, and discovery, and that the sorts of explanatory strategies employed in scientific problem-solving are analogous to the strategies employed elsewhere in human problem-solving. We are concerned primarily with how an explanatory task is seen and how the problem is represented; with how the representation of the problem can influence the character of the resulting theories, and how both are affected by the conceptual and technical tools available in addressing the problems; with what explanatory strategies were adopted, what reasons there were for adopting these strategies, and how explanatory strategies are modified in the face of a new representation of the problem. These are not questions of individual eccentricities or personal quirks; they are applications of problem-solving strategies that have a wide scope, and they are intended as faithful approximations of human cognition. If these explanatory strategies were not common, our own descriptions might be charged with being no more than post hoc descriptions of cases. The fact that they are robustly represented in human cognition, however, constitutes the case for their reality.

The model we will present is itself an approximation, or, more accurately, an idealization—a fragment of a theory of scientific competence. It is a *fragment* because the explanatory strategy on which we will focus is only one among many and is not well suited for all cases. As we will explain in the next chapter, the strategy we will discuss involves *decomposition and localization*—that is, a hierarchical analysis into functional components with specific functions. This strategy is particularly well suited to problems that are relatively ill defined, problems in which neither the criteria for a correct solution nor the means for attaining it are clear. In cases in which the explanatory domain is better defined, alternative strategies may be better suited to the task. There are varying constraints on what constitutes an adequate solution to problems. Experimental methodology and technology are generally tied to these constraints, both by determining what can be answered and by reflecting current theories.

One of the strengths of decomposition and localization as a scientific strategy is that it facilitates an increasingly realistic representation of the explanatory domain, even when the initial representation is seriously distorted: failures of localization can be as revealing as successes. What we offer is a model of scientific *competence* because it is an idealized description, largely abstracting on the one hand from individual eccentricities, and on the other from the vicissitudes of social circumstance. No doubt, both factors will prove essential to any full and descriptively adequate account of scientific change. Galileo's ability to grind lenses surely affected the fate of his theory, as did the need for his expertise in ballistics. Similarly, the opportunities afforded to Pasteur by the French wine industry were no less important to the line of his research than was his role in the French Academy or his expertise with glassware.

We will, thus, focus on *one* constraint on scientific development and discovery. It is a cognitive dimension, but one among many relevant constraints, including a broad range that can be classed as empirical, practical, technological, historical, social, and theoretical. Scientific development, accordingly, becomes a process of simultaneously satisfying a host of constraints. By focusing our attention on one, we will see that others stand out, as if in relief. We will, moreover, discuss only one set of heuristics that might be deployed. As we have suggested, these heuristics are useful when confronted with ill-defined problems in a complex domain.

We propose to use historical cases in detailing the significance of psychological constraints on the development of theories. There are a number of methodological perils in the use of historical case studies, or in what Kevin Kelly and Clark Glymour (1990) derisively call a "retreat to history." One obvious problem is that historical cases can be selected simply because they support some favored theme or set of themes. The use of historical cases then degenerates into elaborate proof-texting and is of little use at all. This problem can be mitigated by using independently plausible and robustly characterized psychological heuristics. For example, Miriam Solomon (1992) has shown that cognitive heuristics proposed and investigated by Tversky and Kahneman can be used to explain a variety of historical features in the development of continental drift in the mid-1900s. Salience, or vividness, of evidence is enhanced if it is concrete, or a matter of personal experience. Such evidence is given a disproportionate weight in comparison to more abstract, and more reliable, statistical evidence. An anecdote from a friend about her car, and the many trips to the shop for its myriad problems, will often outweigh the cold and dry statistical evidence on reliability and repair records provided by *Consumer Reports*. Solomon shows that salience can be used to explain, for example, why scientists with a personal exposure to the biota and geology of the southern hemisphere were more likely to accept drift than were their contemporaries who worked primarily with northern hemisphere data. The same data was available to all. The main differences lie in the way the data was weighted, and this is what Solomon explains in terms of salience, understood as a cognitive heuristic. Solomon does not offer the case study as evidence for the reality of salience. The experimental work of Tversky, Kahneman, and their collaborators has amply demonstrated the role and importance of salience in human cognition. Solomon uses her research, rather, to explain patterns of acceptance in the geological community. By using a mechanism that is independently motivated, Solomon avoids any charge of proof-texting.

We too begin with a heuristic that is independently motivated, though it is lacking the kind of systematic and detailed experimental support provided for salience by Tversky and Kahneman, and we also examine its role in a variety of historical cases. This allows us to explain a number of patterns in the use of evidence and the development of explanatory models, and also provides us with a kind of null hypothesis—a backdrop of expectations against which we may see the operation of other constraints on the development of theories. A parallel might help: It is clear that values affect scientific practice in a variety of ways. The role of social values, or ideological commitments, is particularly evident when theoretical decisions or the assessment of evidence deviates from what would otherwise be normatively appropriate (cf. Richardson 1984). This has led some philosophers to suggest that ideological commitments are present or relevant *only when decisions are otherwise unreasonable*. Larry Laudan, for example, describes this as the *arationality assumption*, the view that sociological and ideological explanations are not necessary when there is an adequate rationalization (1977, pp. 201ff.). This view, however, is mistaken (cf. Longino 1990; Martin 1989): a deviation from what is rationally required *does* make the influence of values more obvious and more readily detectable, but their influence and importance are no less real in cases where they are less obvious. We can use a psychological null hypothesis to detect the presence of other influences on theory development, without holding that these influences are absent unless they are readily detected.[5] To assume that detecting *if* a heuristic is present is tantamount to determining *when* it is present is an egregious and unwarranted abuse of the method.

We thus propose to use historical cases, examining the extent to which we can explain developments and patterns in the history in terms of psychological influences. Through this method we will also see that some developments and patterns cannot be so explained, and we will use such instances as a means for discerning other influences on the development of theories. This will allow us to get some grasp of the range of influences and the importance of psychological constraints on scientific development. We do not retreat to history. We embrace it.

This does not mean we have abandoned the hope for a normative theory of scientific discovery or scientific rationality. Simultaneously psychologizing the account of scientific rationality and insisting on its historical, developmental character does leave us straddling an issue that has loomed large in philosophical thinking about science and philosophy of science. Philosophers of science, like most epistemologists, have presented their endeavors as normative. The task has been thought of as one of prescribing or proscribing what an adequate scientific explanation is, and what is necessary for scientific justification, independent of limitations on decision making or decision makers. The techniques of logical analysis were once supposed to be sufficient to settle these issues. We abandon the a prioristic aspirations and look instead to scientific practice and human reason. We insist on a realistic theory of *human* rationality, and correspondingly on a realistic theory of *scientific* rationality; yet, the enterprise is still, in part at least, normative (cf. Goldman 1986). We propose to reconstitute the normative enterprise rather than reject it.

It is uncontroversial to point out that we cannot simply decide what it is right to do by noting what is actually done. The gap between *is* and *ought* is not this narrow. That apparently leaves us with few options. One is simply to settle for a description of actual practice. We then abandon the search for normative guidelines, or treat normative standards as themselves constituted by particular theories or research traditions. To do this is to yield to a naive historical relativism. Another option is to adopt utopian norms, divorced from the historical episodes, and to ignore the theorist and the tradition in favor of the theory. Just as we may study how clinicians approach medical diagnosis, we may study how scientists approach the task of theory construction and validation; just as we may contrast the practice of clinicians with the formal and analytical actuarial methods that are "normatively appropriate," we may choose to contrast the practice of science with some "normatively appropriate" standard derived from logic or probability theory. Much scientific reasoning—including much of the best—will then violate the norms. To do this is to yield to historical idealism. We seek to forge an alternative between these extremes.

By recasting the normative endeavor to take account of the realities of scientific inquiry and scientific inquirers, we defend a *naturalized* and *humanized* theory of scientific rationality. By describing the strategies scientists use, and simultaneously seeing their limits, we leave room for a normative assessment. We can evaluate actual performance and different human strategies by identifying contexts where they succeed and where they fail, and by seeing which constraints are sufficient and which are not. Such a naturalized theory avoids the Scylla of historical idealism by insisting on standards that are psychologically and developmentally realistic: it is fruitless to prescribe what we cannot perform. Such a theory also avoids the Charybdis of historical relativism by insisting on a distinction between these standards and the actual performance: it is empty to deflate what is right by identifying it with what is done.

2. PROCEDURAL RATIONALITY

Thus, we seek a realistic dynamic model of scientific discovery. We seek to understand the cognitive strategies, the procedures, constitutive of scientific rationality. These strategies are, from one perspective, the procedures that define how humans approach the problem of understanding the world. They define *how* we think about the world. From another perspective, the procedures humans embody constitute assumptions about the structure of the world, or of the part of it to be explained. They define *what* we think about the world.

We take it as axiomatic that rationality is to be understood in terms of problem-solving capacities and skills. Problem solving in general can be understood as a constrained search in a *problem space* that is more or less well defined (cf. Holyoak 1990). In what is by now its canonical form, due to Allen Newell and Herbert Simon (1972), a problem space in a domain includes all possible configurations of the domain, or the number of its possible states. A problem representation, given a problem space, requires four basic components: a *goal state* to be achieved, an *initial state* at which we begin, a set of *operators* defining allowable moves within the state space, and *path constraints* imposing additional limits on what counts as a successful solution. A *solution* is a sequence of operations that leads to the goal state and conforms to the path constraints, and a *problem-solving method* is a procedure for finding a solution for the class of problems at hand. The game of chess provides a useful paradigm for understanding problem solving within a well-defined search space: There are a finite number of possible board positions, each consisting of a permissible arrangement of pieces on the board; these positions define the problem space. Each board position permits only a finite number of alternative moves; these are the path constraints. A solution at any stage would be a series of moves from a given position terminating in checkmate.

This way of understanding problem solving dictates severe practical limits. If a solution requires at least a search through n steps, and if there are M allowable moves at each step, then the number of paths will be M^n. This is literally an astronomical number for even relatively modest values of M and n. If a game of chess involves a total of 60 moves and an average of 30 alternative legal moves at each stage, then the number of paths is 30^{60} or roughly 10^{88} (cf. Newell, Shaw, and Simon 1958). By way of contrast, the universe is something like 15 billion years old. A general brute force solution for chess is obviously out of the question. Some means must be found to limit the search to a computationally tractable number of alternatives. In the face of complex problems, humans typically engage in heuristic search, examining only a small subset of the abstractly possible alternatives. Understanding human problem-solving is, in part at least, understanding the heuristics that guide this search and the way they interact with other variables. To make the point another way, a psychologically realistic model of human problem-solving must respect the limitation imposed by what Simon (1969) calls "bounded rationality": We are organisms limited in memory, attention, and patience. Given these limitations, we limit our search by imposing assumptions about what a solution must look like. These are heuristic assumptions, adopted because of our bounded rationality. Other organisms might impose other assumptions, or other heuristics, but some assumptions are inevitable in the face of complex problems. A psychologically realistic model of human problem-solving must incorporate the heuristic assumptions imposed by human problem-solvers.

Analogously, the task of constructing an explanation for a phenomenon in a given scientific domain is one of finding a sufficient number of variables, the constraints on the values of those variables, and the dynamic laws that are functions on those variables, so that it is possible to predict future states of affairs from descriptions of the universe at an earlier time. A complete state description will pick out a single point in a multidimensional space corresponding to a state of the relevant part of the universe at the time. The number of variables defines the dimensionality of the state space. The dynamic laws are functions from states to subsequent states within this state space. An increase in the number and significance of interactions amounts to an increase in the dimensionality of the problem space needing to be searched. The problem facing a scientific theorist is one of finding laws and variables sufficient to explain what does and does not happen. This is a search in a space of explanatory models, one that will also be subject to a variety of constraints. Some are quite general, such as limitations on human memory. Others are more local, such as limitations on available mathematical models, or simple technological limitations on what data can be gathered. There must therefore be heuristics guiding this search. Understanding scientific problem-solving is, then, a matter of understanding the heuristics that guide the search, both for the relevant variables and the laws.

We also take it as axiomatic that human problem-solving capacities and skills are not domain independent. In some cases, problem-solving methods can be understood as fairly general in application; in others, they are more specialized. Early work in artificial intelligence (AI), including generate-and-test methods, early theorem provers, and Newell and Simon's means-end analysis incorporated methods that could be applied in nearly any domain. They are also weak methods, capable of yielding only relatively poor performance. The moral that the limitations on general search strategies supports is relatively straightforward:

> A system exhibits intelligent understanding and action at a high level of competence primarily because of the specific knowledge that it can bring to bear: the concepts, representations, facts, heuristics, models, and methods of its domain of endeavor. (Feigenbaum 1989, p. 179)

This is by now a generally well accepted moral deriving from work on expert systems and human expertise. Robert Glaser and Michelene Chi explain that, in the light of early work on experts and expert systems,

> it became widely acknowledged that the creation of intelligent programs did not simply require the identification of domain-independent heuristics to guide search through a problem-space; rather, that the search processes must engage a highly organized structure of specific knowledge for problem solving in complex knowledge domains. (Chi, Glaser, and Farr 1988, p. xvi)

We will see that, in the absence of detailed, domain-specific restrictions, the solutions arrived at are rough approximations at best. In Minsky and Papert's terms, a "knowledge-based" strategy is more successful than a "power-based" one (cf. Minsky and Papert 1974, p. 59).[6] An analogous moral extends to human problem-solving. Though some researchers have promoted formal, domain-independent rules (for example, Braine 1978; Rips 1983), it seems reasonably clear that without context-sensitive and domain-specific principles, intelligent problem-solving would be impossible (cf. Cheng and Holyoak 1985, 1989). Rationality should accordingly be understood as requiring the application of specialized cognitive skills and information. Stronger assumptions will limit the search space more, and the more restrictive these assumptions, the more efficient they will be in reaching solutions, if they reach solutions at all; the weaker the assumptions, the more search will be necessary to reach a solution.

Consider the difference between two AI systems, DENDRAL and BACON. DENDRAL is a relatively specialized computer system designed to identify organic molecules from information concerning mass spectrograms and magnetic resonances (Buchanan, Sutherland, and Feigenbaum 1969; Lindsay et al. 1980). After an initial survey identifying selected molecules in a hypothetical sample, DENDRAL generates the set of possible molecular structures on the basis of physical constraints, including such variables as valences and chemical stability. The system then predicts mass spectrograms for these molecular structures and determines a best fit with the data. The generation of possible molecular structures is a procedure allowing for an exhaustive search within the domain defined by the physical constraints. DENDRAL is a model that is highly domain specific, imposing sharp restrictions on the class of solutions, and is very efficient. By contrast, BACON is a program for finding quantitative generalizations and can be applied in a wide variety of fields (see Langley et al. 1987). Given a set of independent and dependent variables, with associated values, BACON looks for correlations that will predict observed values. In its earlier incarnations, BACON would search for constancies and linear relations within the data set and was capable, for example, of inducing Boyle's law from Boyle's own data, Galileo's principle of uniform acceleration, and Ohm's law.

The procedures incorporated in both DENDRAL and BACON are heuristic in motivation and character (for an excellent discussion of heuristics and their role in scientific research programs, see Wimsatt 1980a, 1981, as well as Nickles 1987b), yet they also vary in their power. In some limited cases, in which either the dimensionality of the problem space is low or the problem is simply structured, it is possible to define algorithms that provide an effective and efficient procedure for reaching a solution. Checkers, unlike chess, is a game that lends itself to algorithmic methods: the number of possible board configurations is small in checkers, and the operators are few. In all but the more trivial cases, however, there is no known algorithm that is both effective and efficient for chess or for more complex problems. As a consequence, the most that can be expected is to find procedures that work reasonably well in a limited range of cases. This is accomplished by incorporating assumptions about the domain that limit the number of possibilities to be considered, thereby simplifying the problem-solving task: one limits the search space. In DENDRAL the assumptions about allowable molecular structures limit the search space[7]; in BACON there is a stringent limitation on the form that laws can take, permitting only certain simple, causal relationships to be modeled.[8]

In general, a more domain specific model incorporates assumptions about the domain that are more restrictive and consequently more powerful. For the range of cases defined by such assumptions, the model will be more efficient. On the other hand, a more general model incorporates less restrictive assumptions about the domain, and will therefore be less powerful. In either case, however, the heuristic assumptions imposed are fallible. There will be a range of cases in which a heuristic procedure either will reach no solution at all or will reach an incorrect solution. If we think of the procedure as incorporating a set of assumptions about the task domain, then it will be precisely the cases in which the simplifying assumptions are not met that the procedure will fail. DENDRAL will work only for organic molecules; BACON is limited to laws of a defined set of mathematical forms that, in turn, define the relationships it can see. Thus, not only are heuristic procedures fallible, but their failures are systematic. This dual character of heuristic procedures is an immediate consequence of what makes them both useful and unavoidable; it is the fact that they simplify the problems posed that makes them useful, and it is the simplification that results in systematic failures.

There are parallels to these morals from artificial intelligence in the human case. To take an example that may be relatively familiar, Noam Chomsky has long defended a picture of language acquisition requiring a strong nativist component. The number of possible languages consistent with the sort of data used in learning languages, he claims, is simply too large to be searched effectively without some additional limitations. The nativist component is supposed to specify general features of human languages and thereby limit the search space required to determine the grammar for the language being learned. Chomsky understands language acquisition, in part, in terms of restrictions imposed by this nativist component. At the same time, the nativist component incorporates nontrivial assumptions about the structure of human language, thus imposing assumptions about the structure of the language to be learned. If such nativist procedures are used in learning language, then there exist languages we simply cannot learn precisely because their structures violate the assumptions built into these heuristics.

Our problem in approaching a model for scientific discovery is structurally analogous to work on problem solving. Scientific discovery involves a heuristically guided search for solutions in a complex problem space. However, unlike the games that provide the prototype for problem solving, scientific problems are often ill defined; that is, the constraints defining an adequate solution are not sharply delineated, and even the structure of the problem space itself is unclear (cf. Reitman 1965). As Simon (1973b) points out, an important part of real world problem-solving is, often enough, imposing an appropriate structure on the problem. But, whether well or ill defined, there must be some means for restricting the relevant search space. One way to do this is by imposing some restrictions concerning possible solutions. We may limit the number of variables, assuming only some will make a significant difference. For example, if we are interested in predicting the orbit of a planet, the influences of other planets, for the most part, induce only minor perturbations; however, for other purposes, such as assessing the influence of orbital variations on global climate, their influence is more significant. Alternately, we may impose assumptions about the form of the relevant laws. Thus, in population dynamics it is usually acceptable to assume that the functions will be linear and influences will be additive, although this is not always true. As

Robert May (1974, 1976) initially pointed out, even relatively simple systems will exhibit complex dynamic properties given appropriate values for population size and variables such as birthrate. Limiting the relevant variables and imposing assumptions about the form of relevant laws are procedures for attacking problems. They also describe partial solutions. Applied to scientific problem-solving, this would mean that the heuristic assumptions constitutive of explanatory models would be critical in a developing research program. Without them, it would not be possible to formulate or develop a practical plan of research. The need for heuristic assumptions also raises the prospect that a program of research is misguided, its results simply artifacts of the simplifying assumptions that define the program.

The unification of the procedural and the descriptive will engender philosophical resistance, but it is a natural consequence of a shift away from abstract models of rationality to ones that are sensitive to a demand for realism. If rationality is to be understood in terms of problem-solving capacities and skills, then insofar as human problem-solving capacities and skills are specialized to particular task domains, rationality should be understood as a matter of the application of specialized abilities. This application, in turn, is a matter of incorporating assumptions about the structure of the domain under investigation.

CHAPTER 2

Complex Systems and Mechanistic Explanations

1. MECHANISTIC EXPLANATION

Our aim is to develop a cognitive model of the dynamics of scientific theorizing that is grounded in actual scientific practice. Our focus is on one kind of explanation, one involved in understanding the behavior of complex systems in biology and psychology. Examples of the complex systems we have in mind are the physiological system in yeast that is responsible for alcoholic fermentation, and the psychological system responsible for memory of spatial locations. As we shall discuss in this section, these explanations, which we refer to as *mechanistic explanations*, propose to account for the behavior of a system in terms of the functions performed by its parts and the interactions between these parts. The heuristics of decomposition and localization are central to our analysis of the development of mechanistic explanations. We shall discuss these heuristics in some detail in this chapter, especially in sections 2 and 3 of this chapter. Parts II and III will be concerned with illustrating their role, and also their development under other influences. These heuristics, or family of heuristics, can be thought of as imposing assumptions about the organization of the systems being explained. We shall examine these assumptions and the kinds of organization of actual systems for which these assumptions are likely to succeed and those on which they will likely fail.

By calling the explanations *mechanistic*, we are highlighting the fact that they treat the systems as producing a certain behavior in a manner analogous to that of machines developed through human technology. A machine is a composite of interrelated parts, each performing its own functions, that are combined in such a way that each contributes to producing a behavior of the system. A mechanistic explanation identifies these parts and their organization, showing how the behavior of the machine is a consequence of the parts and their organization. What counts as mechanistic, though, changes with social context. Scientists will appeal analogically to the principles they know to be operative in artificial contrivances as well as in natural systems that are already adequately understood. The state of technology and natural science at any given time thus plays a significant role in determining the plausibility and limits of mechanistic explanations (cf. Gregory 1981). From the universe of the *Timaeus*, through the Archimedian analogues of Galileo and the clockwork universe of Newton, to the recent focus on servo-mechanisms and computers, the available analogues were important factors in determining which mechanistic models scientists advanced.

The nature and plausibility of mechanistic models is also influenced by characteristics of human thinking, especially by the proclivity of humans to trace operations in a linear or step-by-step fashion. This is especially evident when we consider the forms of organization possible. Many machines are simple, consisting of only a handful of parts that interact minimally or in a linear way. In these machines we can trace and describe the events occurring straightforwardly, relating first what is done by one component, then how this affects the next. Such machines induce little cognitive strain. Some machines, however, are much more complex: one component may affect and be affected by several others, with a cascading effect; or there may be significant feedback from "later" to "earlier" stages. In the latter case, what is functionally dependent becomes unclear. *Inter*action among components becomes critical. Mechanisms of this latter kind are *complex systems*. In the extreme they are *integrated systems*. In such cases, attempting to understand the operation of the entire machine by following the activities in each component in a brute force manner is liable to be futile.

A major part of developing a mechanistic explanation is simply to determine what the components of a system are and what they do. In broad outline, there are two strategies available to analyze and isolate component functions. The first is to isolate components physically within the system and then determine what each does (the goal is to use the knowledge of components to reconstruct how the system as a whole operates). The second strategy is to conjecture how the behavior of the system might be performed by a set of component operations, and then to identify components within the system responsible for the several subtasks. The former is the *analytic* method of Etienne Condillac, which played a role in the development and promotion of both Lavoisier's chemistry and Cuvier's functional anatomy. The latter is the explanatory program of functionalism in contemporary philosophy of mind (see Cummins 1983; Dennett 1978; Lycan 1981a, b), which for contrast we will refer to as a *synthetic* strategy (cf. Posner and McLeod 1982). An analytic strategy constructs from the bottom up; a synthetic strategy projects from the top down.

The analytic strategy is confronted by the fact that smoothly operating systems conceal their component operations. As we will see repeatedly in ensuing chapters, the breakdown of normal functioning often provides better insight into the mechanisms than does normal functioning. In the absence of natural error, failure can be induced. Simon develops the point clearly:

A bridge, under its usual conditions of service, behaves simply as a relatively smooth level surface on which vehicles can move. Only when it has been over-loaded do we learn the physical properties of the materials from which it is built. (1981, p. 17)

Although overtaxing a system induces malfunction, which can be an important clue to the functional properties of component parts, this is a relatively crude method—one as likely to lead to catastrophic breakdown and chaos as to insight. For that reason, researchers prefer to differentiate components and their functions by altering specific activities *within* a system. Such *inhibitory* or *deficit studies* allow us to determine a physical component's contributions to the system by inhibiting its operations and then observing resulting deficits in overall system behavior. The best-known examples of such studies are ablation studies in nineteenth-century physiology (see Harrington 1987), but chemical poisoning studies in biochemistry, and the use of x-rays in the genetic research of Beadle and Tatum, follow the same logic.

Inhibitory studies are problematic for a number of reasons. Perhaps the most serious danger is the temptation to infer, from the fact that a specific experimental manipulation interrupted or inhibited a particular activity of the system, that the part of the system damaged is the component responsible for that activity. As R. L. Gregory pointed out over thirty years ago,

> Although the effects of a particular type of ablation may be specific and repeatable, it does not follow that the causal connection is simple, or even that the region affected would, if we knew more, be regarded as functionally important for the output—such as memory or speech—which is observed to be upset. It could be the case that some important part of the mechanism subserving the behavior is upset by the damage although it is at most indirectly related, and it is just this which makes the discovery of a fault in a complex machine so difficult. (1961, p. 323)

Simplistic uses of functional deficit studies might lead us, for example, to conclude that a resistor in a radio is a hum-suppressor because the radio hums when the resistor is removed (Gregory 1968, p. 99).

Localization based on deficit studies is often erroneous. What is required is some means of figuring out, from the observed deficits, what the component in question positively contributes to the system when it is functioning normally. These are the functions that a good mechanistic explanation will localize in the parts. We will see that an important guide to finding these functions is the simultaneous use of a variety of inhibitory techniques. This can sometimes allow the investigator to determine component functions, but such an account will be complete and compelling only when there is an explanation of how these components interact in effecting the normal operation of the system.

An alternative analytical technique follows an opposite approach, stimulating a physical component and observing the behavioral effects on the whole system. If extra stimulation produces an identifiable surplus, we can sometimes infer that under normal conditions the component was responsible for that which is now generated in excess. We shall generally refer to such investigations as *excitatory studies*. The best-known examples are stimulation studies in neuroscience in which electrical stimulation is applied directly to the cortex. Biochemical studies in which potential metabolic intermediaries are injected exhibit a similar logic. Once again, however, there is a temptation to infer that, because excitation to a physical system enhances or induces a particular effect, the stimulated part is necessarily the seat of the function in question. With both inhibitory and excitatory studies, the natural operation of the system is modified; the experimental techniques may be the source of experimental artifacts and may not be diagnostic of normal operations within the system. They are, however, the only techniques available in some cases. Even lacking a clear conception of the organization of the system, analytical strategies allow the experimentalist to probe the system and its organization.

A synthetic strategy requires some prior hypothesis about the organization and operation of the system. From an initial hypothesis about the underlying mechanisms, one formulates a model of how the system functions. The empirical task involves testing performance projected on the model against the actual behavior of the system. One discipline using model studies is AI, in which researchers propose and implement a set of operations in order to perform an activity that would require intelligence if performed by a human. Such model studies are also used in other domains. In a later section we will discuss biochemical research in which comprehensive proposals were advanced as to the intermediate chemical reactions in an overall physiological process. This theoretical work was sometimes supported by actual development of artificial systems intended to show that a mechanism such as the one being proposed could in fact carry out the needed process.

As plausible as such a model might be, it might also turn out to be radically misconceived. One such case is Justus Liebig's (1842) general account of nutrition. Liebig proposed a model of nutrition which plotted a complete set of metabolic transformations, based only on information about the chemical composition of food and waste products. He was guided by the assumption that because food substances are more complex than waste products, all that an animal does is break down these substances to release the energy stored in them. Using his knowledge of basic chemical reactions, Liebig proposed a bold and brilliant model of animal metabolism. It was also wrong. Though consistent with the data Liebig used, it foundered on physiological data introduced by Bernard, who later showed that animals not only break down complex foodstuff, but also synthesize substances necessary for life (see Bechtel 1982).[1]

The synthetic strategy is traditionally regarded as speculative. Perhaps it is. Speculative stages are nonetheless important to all science, if only in identifying possible mechanisms in contexts in which no workable mechanisms were envisioned. The subsequent testing of these models can and does induce the development and elaboration of more adequate models. However, speculation without empirical constraints is as likely to produce spurious explanations as correct ones.[2] In some cases, one might posit component operations that are, in fact, as complex as the overall operation. We will then have no net explanatory gain. In other cases, one might propose only one among many possible mechanisms and have no warrant for thinking it is the one actually utilized. This is hardly better than having no explanation at all.

The analytic and synthetic strategies are complementary. Inhibitory and excitatory studies can provide empirical data appropriate for evaluating synthetic models. Moreover, synthetic models can provide a theoretical framework in which to interpret information obtained from the empirical studies employing analytical strategies. Even together, though, the strategies hardly provide a fail-safe methodology. The dangers of spurious explanation and premature localization still confront the scientist.

Several factors can make the process of developing mechanistic explanations using techniques such as these extremely challenging. To begin with, as the level and significance of interactions increases, the complexity of the explanatory problems increases as well. The task of constructing an explanation for a given domain might be viewed as one of finding a sufficient number of variables, the constraints on the values of those variables, and the dynamic laws that are functions of those variables. These laws make it possible to use the model to predict future states of affairs from descriptions of an earlier time. The number of variables, once again, defines the dimensionality of the state space or problem space. A complete state description will pick out a single point in a multidimensional space corresponding to a specific value for each variable in the relevant part of the universe at the time. The dynamic laws are functions from states to subsequent states within this state space.

Let us focus just on the question of how complex the state space will be that we need to consider in order to represent the state of the domain. An increase in the number and significance of interactions posited in the constraints and laws requires an increase in the dimensionality of the state space needed to represent the domain. An example may help. Richard

Lewontin (1974) explains that in the case of genetics we can think of the problem as one of predicting the genetic composition of a population from its composition at an earlier time. As we consider greater numbers of loci and alleles, the number of variables needing to be considered increases and, hence, so does the dimensionality of the state space. What is pertinent to our purposes is how interaction affects the number of variables, and consequently the number of dimensions in the state space. If we assume that genes at different loci segregate independently, then the predictive problem reduces to one of finding solutions to independent problems for each locus, and accordingly the proper unit of analysis will be allelic frequencies. In that case, the dimensionality of the space needed to represent the state of the genome is a linear function of the number of alleles and the number of loci. With n loci and a alleles at each locus, the dimensionality of the problem space approximates the product of the two. If, on the other hand, we assume that there is significant linkage between genes at different loci, then, Lewontin urges, the proper unit of analysis will be gametic frequencies. As the number of loci increases, the number of possible gametes increases exponentially: in the case of two alleles at each locus, with two loci, there will be four gametic classes; with three loci, there will be eight gametic classes; and with ten, there will be over a thousand. The expected dimensionality for the state space will be one less that the number of allelic or gametic classes. This means that the anticipated dimensionality of the state space needed to represent the state of the genome will also increase linearly with allelic classes as the unit, and exponentially when using gametic frequencies. The results are detailed in Table 2.1.

Explanatory demands are further aggravated by the relative stability of the systems involved. This is part of what helps to conceal the contributions of the individual components when we examine a normally operating system. As we noted previously, if the goal is to reveal what the parts contribute to the operation of the whole system, we generally must find techniques to perturb the system. When investigating the behavior of *self-organizing* systems, the theorist must contend with the fact that these systems will maintain or determine an equilibrium even in the face of considerable perturbations, both internal and external. Consequently, establishing what the parts contribute will be difficult. In the limiting case, self-organizing systems will be homeostatic—that is, they will maintain a predetermined equilibrium state despite perturbations. The genetic system, in which endonucleases repair damage to nuclear DNA, is one example of such a homeostatic system. In more complicated cases, the systems are better thought of as *self-regulating*—that is, they will modulate activity at varying levels depending on other influences. For example, coenzymes in cell metabolism adapt the breakdown of foodstuff to the work performed by the cell.

n	$a = 2$		$a = 5$		$a = 10$	
1	1	**1**	4	**4**	9	**10^{+1}**
2	2	**3**	8	**24**	18	**10^2**
3	3	**7**	12	**124**	27	**10^3**
4	4	**15**	16	**624**	36	**10^4**
5	5	**31**	20	**3,124**	45	**10^5**
6	6	**63**	24	**15,624**	56	**10^6**

Table 2.1. Increase in the Dimensionality of a Problem in Genetics as a Function of Linkage. With n loci and a alleles at each locus, the dimensionality of the problem depends critically on linkage. Assuming independent assortment (indicated in plain text), the proper unit of analysis will be allelic frequencies. The dimensionality of the state space increases as a linear function of n. With linkage effects (indicated in bold text), the proper unit of analysis is gametic frequencies. The dimensionality then increases exponentially. (Based on Lewontin 1974.)

When a system's behavior is relatively constant despite variations external to it, we can safely ignore the environment and focus only on the internal mechanisms to specify the parts, their interactions, and their contributions to the behavior of the system. But when the system adapts to the environment, as homeostatic and self-regulating systems do, we cannot simplify the task in this way. The sensitivity to environmental changes means that the parts operate differently under altered conditions and so further conceal from view how they behave when the system is operating normally. In general, interaction among the various components makes it difficult to isolate independent contributions from their coordinated output.

2. DECOMPOSITION AND LOCALIZATION

We now turn to the heuristic strategies of decomposition and localization. As we will see in the chapters to follow, these strategies have been used by a wide variety of researchers in a wide variety of disciplines, from nineteenth-century brain science and early twentieth-century investigations into chromosome structure to more recent work on language and cognition.

Decomposition allows the subdivision of the explanatory task so that the task becomes manageable and the system intelligible. Decomposition assumes that one activity of a whole system is the product of a set of subordinate functions performed in the system. It assumes that there are but a small number of such functions that together result in the behavior we are studying, and that they are minimally interactive. We start with the assumption that interaction can be handled additively or perhaps linearly.

Whether these assumptions are realistic or not is an open question; indeed, at the outset we often simply do not know. The extent to which the assumption of decomposability is realistic can be decided only a posteriori, by seeing how closely we can approximate system behavior by assuming it. We may be led to erroneous explanations, but it may be the only way to begin the task of explaining and understanding complex systems. The failure of decomposition is often more enlightening than is its success: it leads to the discovery of additional important influences on behavior.

Localization is the identification of the different activities proposed in a task decomposition with the behavior or capacities of specific components. In some cases we may be able to identify (through fairly direct means) the physical parts of the system in which we can localize different component functions. In other cases we may have to rely on various functional tools for determining that there are such parts, without being able to identify them; for example, we may be able selectively to inhibit their operation and observe the consequences on behavior. We need not assume that a single part in this sense is a spatially contiguous unit; in fact, we know that in many cases it is not. A functional unit may be distributed spatially within the system. Localization does entail a realistic commitment to the functions isolated in the task decomposition and the use of appropriate techniques to show that *something* is performing each of these functions.

In one extreme form, decomposition and localization assume that a single component within the system is responsible for some range of phenomena exhibited by the system. For example, it is assumed that the posterior cerebral lobes are responsible for vision, that the cell nucleus is responsible for genetic control, and that there is a specialized enzyme responsible for catalyzing a given chemical reaction within the cell. This, the simplest assumption, often guides the first explanatory models, even if it seldom survives. We refer to it as *simple* or *direct localization*, being simple both in focusing on single components and in imposing the fewest constraints. The simplest case, however, is often far too simple (cf. Ch. 4). The behavior to be explained may be at best the product of several independent components rather than of one. No one component can be assigned sole responsibility. It also may be necessary to assume there is some interaction and some differentiation of function. Lacking simple localization, the alternative is to localize a set of component functions and assume linear interaction will explain the behavior of the system. We refer to this as *complex* or *indirect localization*, having not only a complex organization, but complex constraints on the problem.

Pursuing decomposition and localization is to impose an assumption about the nature of the system whose activities one is trying to explain: it is assuming that it is *decomposable*. A decomposable system is modular in character, with each component operating primarily according to its own intrinsically determined principles. Thus, each component is dependent at most upon inputs from other components, influences other components only by its outputs, and has a specific, intrinsic function. The notion of decomposability stems from Simon and constitutes the descriptive counterpart of localization. Localization presupposes that we are confronted with a modular organization such that the components of the system can be subjected to separate study and investigation; it requires that the components have discrete intrinsic functions intelligible in isolation, even if such functions do not independently replicate those of the system as a whole.

An extreme form of a decomposable system is an *aggregative system*,[3] which is a species of *simply decomposable* system. System behavior in such cases is a linear or aggregative function of component behavior (cf. Levins 1970, p. 76). Whatever organization is present is not a significant determinant of the relevant systemic properties. In an important paper, "Forms of Aggregativity," Wimsatt lays down four conditions of aggregativity (1986a, pp. 260–68):

1. Intersubstitutability of parts;
2. Qualitative similarity with a change in the number of parts;
3. Stability under reaggregation of parts; and
4. Minimal interactions among parts.

We emphasize the last condition, primarily because the focus of our investigation is the discovery of organizational properties that fix the interaction of the parts and determine their significance for system behavior. However, we do not see a useful way to elaborate on this condition independently of the first three, and we think that—with a suitable decomposition—when this last condition is satisfied the other three generally will be satisfied as well.[4] In offering the final condition, Wimsatt gives us this statement, with some reservations: "There are no cooperative or inhibitory interactions among parts of the system" (ibid., p. 269). As an example, lateral inhibition in the retina leads to a lower activation among neurons than would be predicted if we attended only to their stimulation level; this is because an elevated activation level for one neuron decreases the activation level of adjacent neurons.

Few interesting dynamic systems are strictly aggregative. For example, fluid flow or the movement of a herd *approximates* aggregative motion, though even these *only* approximate aggregative systems and some of the most interesting work concerns why this is so. When the relevant systemic properties are at least partially determined by the organization of the system, we no longer have aggregativity. Or, more realistically, to the extent that organization determines systemic behavior, the system is nonaggregative. In the simplest departures from aggregativity, we may still maintain intersubstitutability; however, when this also fails, we have what we call *composite systems*. There are two species of composite systems, which differ in terms of the role played by systemic organization: In *component systems*, the behavior of the parts is intrinsically determined. In these cases it is feasible to determine component properties in isolation from other components, despite the fact that they interact. The organization of the system is critical for the functioning of the system as a whole, but provides only secondary constraints on the functioning of constituents. In *integrated systems*, systemic organization is significantly involved in determining constituent functions. There may be, for example, mutual correction among subsystems, or feedback relations that are integral to constituent functioning. Thus, as we will see, although work on cell metabolism by, for example, Neuberg and Thunberg in the early decades of this century treated metabolic processes as linear and sequential, the discovery of coenzymes and their function has made it clear that linear models are oversimplified in the extreme (see Bechtel 1986a, and Chapter 7). Systemic organization then provides primary constraints on constituent functioning, and constituent functioning is no longer intrinsically determined. Some of the most interesting and bewildering problems arise with such systems. Richard Levins comments:

> This is a [type of] system in which the component subsystems have evolved together, and are not even obviously separable; in which case it may be conceptually difficult to decide what are the really relevant component subsystems. Thus, for example, we might consider that a simpler multicellular organism is composed of cells, and yet the cells may be more profitably regarded, under other circumstances, as simply spatial subdivisions, partly isolated, of an organism. (1970, p. 77)

Dependence of components on each other is frequently mutual and may wholly blur any distinction among them. Thus, mitochondria were once independent organisms—though they are now clearly but parts of a cell, integrated into cell metabolism (Margulis 1970), and we cannot now understand how they function if we neglect their incorporation in, and integration into, the complex activities of the cell.

Composite and integrated component systems correspond to two types of organization in Simon's scheme (component systems are Simon's *nearly decomposable* systems). To the extent that components perform independent functions and send their outputs to other parts, we have *strict* or *simple decomposability*. *Near decomposability* imposes less stringent limits, as Simon explains:

> (1) In a nearly decomposable system, the short-run behavior of each of the component subsystems is approximately independent of the short-run behavior of

the other components; (2) in the long run the behavior of any one of the components depends in only an aggregate way on the behavior of the other components. (1969, p. 210)

For near decomposability, individual components must be controlled by intrinsic factors, in the manner of composite systems. As components become less governed by intrinsic factors, we enter the domain of integrated composite systems, which are *minimally decomposable.*

A system will be nearly decomposable to the extent that the causal interactions *within* subsystems are more important in determining component properties than are the causal interactions *between* subsystems (cf. Simon 1969, p. 209). Wimsatt (1972, p. 72) suggests that we characterize such systems in terms of a parameter that is a "measure of the relative magnitudes of intra- and inter-systemic interactions for these subsystems." As we noted above, the heuristics of decomposition and localization assume a degree of decomposability. At a minimum they assume that the system is at least nearly decomposable. A parameter of the sort Wimsatt offers, then, would be an estimate of the likely error in predictions based on models developed using decomposition and localization (for further discussion, see Richardson 1982).

Whether decomposition and localization will succeed or fail in a given case, these heuristics are important because they provide us with a tractable strategy for attacking the explanatory problems complex systems present. Recent research in the psychology of judgment indicates that humans have great difficulty comprehending cases with more than a few interacting variables. Humans *cannot* use information involving large numbers of components or complex interactions of components, and even when the problem tasks are computationally tractable, human beings *do not* approach them in this way. Complex systems are computationally as well as psychologically unmanageable for humans.

3. Hierarchy and Organization

We need now to explore possible reasons for thinking that the sorts of systems encountered in biology or psychology are likely to be decomposable or nearly decomposable—and hence amenable to mechanistic explanations developed through decomposition and localization—or only minimally decomposable or not decomposable at all—and hence not amenable to explanation using these heuristics. To explore this issue it is useful first to recognize that in talking about components and whole systems we are construing nature as incorporating a hierarchy of levels. When parts interact with each other, we can view this as a *horizontal* process; that is, there is interaction between units at one level. When we focus on how components combine into larger units, which in turn may interact with other larger units, we are addressing a *vertical* question about the relations between levels. If the degree of interaction among components when they come to form wholes does not obliterate the components as autonomous entities, the result is a *decomposable part-whole hierarchy*.

A hierarchical organization also facilitates tractability by providing for theoretical economy. In explaining the behavior of the system we can gain independent characterizations of each component, ignoring both the contributions of other components at the same level as well as the influences operative at higher or lower levels. For example, programing in higher-level languages allows us to bypass the means by which the commands are executed. If the system is decomposable, there will be relatively little information lost in such a representation: "In studying the interaction of two large molecules, generally we do not need to consider in detail the interactions of nuclei of the atoms belonging to the one molecule with the nuclei of the atoms belonging to the other" (Simon 1969, p. 218).

Moreover, there is considerable evidence that information is held in human memory within organized structures and that the very intelligibility of a domain may depend on its being represented as a decomposable hierarchy. In work by Simon and his collaborators, it has been shown that experts in various disciplines differ from novices not just in the amount they know, but in the way their knowledge is organized. Expert chess players, for example, are readily able to reconstruct board positions of games that they have observed for only a few seconds, whereas novices are able to locate only a few pieces. This difference disappears, however, when the board positions do not make strategic sense. Chase and Simon (1973a, b) contend that the differences are due to the fact that experts recognize and remember patterns among pieces and treat these patterns as units (see also Gilmartin and Simon 1973). Analogously, in tasks evaluating the efficiency of recall and recovery, memory is facilitated by texts with specific forms of organization and inhibited by others. Free recall will even impose organization when none is present in the text.

Simon holds that hierarchical organization is also a general phenomenon in nature. He argues that hierarchies arise because the forces governing interactions between objects typically do not form an equally distributed continuum. The strongest forces govern interactions at the lowest level and give rise to reasonably stable units at a middle level. In many cases these lower-level forces may not determine which among a variety of complexes at the higher level will be realized. They constrain but do not determine the results. Weaker forces then come to play in determining the relationships between these middle-level entities.

As an illustration, Simon considers chemical forces. The forces responsible for atomic structure are stronger than those that determine the composition of the molecules made of them. Similarly, the forces determining the structure of macromolecules are weaker than those that determine their composition.

> Thus, protons and neutrons of the atomic nucleus interact strongly through the pion fields, which dispose of energies of some 140 million electron volts each. The covalent bonds that hold molecules together, on the other hand, involve energies only on the order of 5 electron volts. And the bonds that account for the tertiary structure of large macromolecules, hence for their biological activity, involve energies another order of magnitude smaller—around one-half of an electron volt. It is precisely this sharp gradation in bond strengths at successive levels that causes the system to appear hierarchic and to behave so. (Simon 1973a, p. 9)

An equilibrium between the forces at the lowest level defines a set of stable systems. The forces at the higher level then determine the relationship between these units and their combination into other units. In biochemical processes it has been clear since the work of Linus Pauling and Max Delbrück (1940) that quantum mechanical forces are insufficient to explain reactions among complex molecules in the cell; much weaker forces are required for the intermolecular interactions. Hydrogen bonds are especially important, both in antibody formation and in the classic work on the double helix by Watson and Crick (1953). The resulting structures again form into stable units, with their interrelations defined by still weaker forces.

The argument so far for nature comprising decomposable hierarchies assumes that the strengths of the forces for binding components into structures are not continuously distributed. This may or may not be true, but Simon offers other, more general, arguments. For example, he appeals to evolutionary considerations, arguing that complex systems are more likely to evolve if they are hierarchical and decomposable. He assumes that the lower-level forces insure that components can arise independently, existing as stable units in their own right. All that would then be necessary is the formation of stable combinations that would meet the demands at the higher level. Given complex macromolecules, living cells would be combinations formed from them, dependent only on the operation of higher-level forces. Selection could then serve to fine-tune the system without large-scale disruption. To quote once again from Simon,

> The loose horizontal coupling of the components of hierarchic systems has great importance for evolutionary processes just as the loose vertical coupling does. The loose vertical coupling permits the stable subassemblies to be treated as simple givens, whose dynamic behavior is irrelevant to assembling the larger structures, only their equilibrium properties affecting system behavior at the higher levels.

The loose horizontal coupling permits each subassembly to operate dynamically in independence of the detail of others; only the inputs it requires and the outputs it produces are relevant for the larger aspects of system behavior. (1973a, p. 16)

To illustrate the principle, Simon (1969) tells the following tale of two watchmakers: Each makes fine watches with 1000 parts of diverse sorts. One of the two, Tempus, uses an hierarchical design with component parts: each watch has ten components, and each component has ten components, etc. The other, Hora, uses a horizontal design. All the parts must be in place before the watch will stay together. Each of the watchmakers is interrupted periodically to take orders for additional watches. Hora's work suffers dramatically, because every interruption results in a loss of all the work on the current watch. Tempus' work suffers too, though not so dramatically, because all that is lost is the work on the current subassembly. The moral is a general one: complex structures are more efficiently constructed if they are composed of stable subassemblies. Simon shows that additional levels of intermediate structure will further increase stability; consequently, the time required to assemble a system with any given number of units is inversely proportional to the number of intervening levels. He then goes on to apply the same principle to the evolution of biological units: decomposability increases the evolutionary rate (ibid., pp. 200ff.). As a result, complex systems arise more readily when they consist of stable subsystems.

As we noted in Chapter 1, the assumption that a system is nearly decomposable and hierarchical is not just motivated by theoretical arguments; human cognitive strategies make such an assumption natural. This does not mean, however, that it is realistic. There are in fact reasons to suspect that many natural systems are not decomposable even though they are hierarchical. Simon concentrates on the division of systems into component parts. In minimizing interconnections between these parts, and treating them as autonomous, Simon sidesteps discussion of what binds parts together, making them *parts* of a complex system. If nothing imposes systemic structure, we have an aggregative system. Systemic behavior is an additive function of component behavior, at least if we ignore threshold effects. We have a hierarchy in name only. With composite systems, interaction between units is critical; indeed, it is constitutive of the higher-level units. Interaction is what makes composites useful for explanatory purposes; however, interaction also compromises the autonomy of components.

The mode of organization is important. Grobstein (1973) distinguishes between *facultative* and *obligate* organization, in a transparent analogy to social symbioses (cf. Richardson 1982). A facultative organization allows members to disperse and recombine. Individuals can function as part of a more complex system, but are also capable of independent activity; for example, baboons will often forage in groups, but they are also able to forage independently. Facultative organization is thus nearly decomposable. An obligate organization, by contrast, is one in which interdependence has significantly compromised the capacity for independent activity, and the system is thus only minimally decomposable. As Grobstein says, the properties of higher-level structures "are in some sense immanent in the properties of the components, [though] . . . such properties tend to be lost if components of a set are dispersed or if a set is dissociated from its context or superset" (1973, p. 45). Some flowering plants are wholly dependent on birds or insects as vectors for pollination; and some of these birds and insects, in turn, are specialized, feeding on only one type, or a few types, of flowering plants. The latter form of organization is particularly important, as Grobstein recognizes:

> Their components are very different in properties when in isolation or in the collective, and the collective, once formed, is not reversible. This is the case with most higher organisms. . . . A complex multicellular organism represents an extreme case in which very special conditions are required to maintain individual cells or individual organs outside of the collective relationship. (Ibid., p. 34)

Not only are the relationships among constituents of the system important for explaining the system's operation, but the constituents themselves have no independent, isolable function.

Simon's theoretical arguments supporting the ubiquity of nearly decomposable systems with facultative organization rest on evolutionary considerations. Such considerations, though, will not support the conclusion that complex systems are generally decomposable. Levins (1970) and Wimsatt (1972) point out that divergence and coadaptation will decrease the decomposability of a system with time: once aggregated, components can diverge and specialize in functions while maintaining stability. Evolvability does not insure stability; thus, while decomposable hierarchies may be more likely to evolve, the considerations advanced do not necessarily favor maintaining decomposability once they are formed. It will remain an open question whether complex natural systems are necessarily decomposable hierarchies.

4. CONCLUSION: FAILURE OF LOCALIZATION

Whether natural systems are hierarchically structured will influence how successful decomposition and localization will be as heuristic strategies. In this chapter we have described decomposition and localization as heuristics for developing mechanistic explanations of complex systems and have examined the assumptions these heuristics make about the nature of the system we need to explain. While there are considerations favoring the occurrence of decomposable hierarchies, there are also considerations pointing toward only minimally decomposable, integrated systems. Thus, there are clearly risks in assuming complex natural systems are hierarchical and decomposable. There are always some risks that stem not from the assumption of decomposable hierarchies, but from specific errors in developing the decomposition and localization. In a case of false localization, a complex system may manifest a component organization, but not the specific component analysis attributed to it. In this case the way the system operates is misrepresented. This may be because there is an alternative component organization at the same level, or because we have adopted the wrong level of analysis altogether. We will consider cases subsequently in which the initial analysis is misguided in these ways.

More radical errors are also possible. The separation of systems into isolated components, with the attendant minimization of interactive importance, may blind us to critical factors governing system behavior; in particular, it may blind us to the importance of systemic interaction. We may not have a decomposable system at all, or we may have one that is only minimally decomposable. Simon acknowledges the risk saying, "If there are important systems in the world that are complex without being hierarchic, they may to a considerable extent escape our observation and understanding" (1969, p. 219). The risk, if realized, should be felt in failures of explanation. If the failures are more limited, we are only limited in our explanations. We will also examine failures of a more radical sort in which, though research began with the assumption of near decomposability, high degrees of organization were subsequently recognized and the explanatory approach had to be adapted to accommodate this organization. Though decomposability may be a natural and fruitful starting point, it may be no more than that.

PART II

Emerging Mechanisms

It has been found in science that when a sub-universe of discourse can be dissociated from a larger universe and a means of studying behavior found which is but slightly affected by uncontrollable factors, the results usually have a high value in prediction.

—E. M. East 1934

Every organized being forms a whole, a unique and closed system, whose parts mutually correspond and concur to the same definite action by a reciprocal reaction. None of its parts can change without the others also changing; and consequently each of them taken separately, indicates and determines all the others.

—G. Cuvier 1812

INTRODUCTION

In Chapter 1 we emphasized that our investigation is directed at identifying the cognitive constraints affecting theory development, and in Chapter 2 we introduced decomposition and localization as the central heuristics figuring in our treatment of the development of mechanistic explanations. We turn now to developing a more detailed analysis, one grounded in historical analyses, of how scientists actually develop mechanistic models. As we proceed, we will focus on *choice-points*, points at which decisions are made that shape the explanatory endeavor. The decisions scientists make are affected by their own cognitive characteristics (for example, the fact that they are agents with bounded rationality); theoretical considerations that suggest that one or another sort of explanation might be viable for the particular problem; and available empirical data. In this part we will identify some of the initial choice-points that are confronted before actually developing an explanation that would qualify as fully mechanistic. These focus on the identification of discrete systems in nature, the assignment of activities to them, and the determination of whether these systems can be functionally decomposed.

Before it is possible, or even relevant, to develop a fully mechanistic explanation of *how* a system performs some function—and, therefore, before the heuristics of decomposition and localization are properly brought into play—it is necessary to identify *what* functions are preformed and *what* system performs these functions. We speak of this as isolating the *locus of control*. A locus of control is not necessarily a system that operates in isolation; rather it is one that carries out a transformation of inputs into outputs that *is* what constitutes realizing a specific function. Although such an identification of a locus of control is critical to any attempt to develop a mechanistic explanation, claims to have identified such loci are often controversial.

We start in Chapter 3 with a discussion of two domains in which controversy is still alive. One involves identifying the locus of control for behavior. The other involves the locus of control in evolution. In each case there are prominent scientific traditions placing control in the environment. Radical behaviorists argue that organisms are not loci of control for behavior—that such control lies outside the organism. Likewise, the Darwinian explanation of evolutionary adaptations looks to the forces of selection operating on individuals of the species, rather than to factors internal to such individuals. Each of these positions can be placed in counterpoint to another which sees the system as, in an important sense, serving as a locus of control. Cognitive psychology, as a recent mentalist turn, rejects the claim that one can understand behavior without looking inside the organism, and so treats the cognitive system itself as the locus of control. A similar argument has been made in the case of evolutionary theory. This is particularly true of Haeckel's and other orthogenetic programs in the nineteenth century, which view evolution as an internally directed process leading to increasingly complicated forms of organization.

There are other cases in which such controversies have been resolved. One was the nineteenth-century controversy over the locus of control for respiration. Although experimental evidence was marshaled, this alone was not sufficient to settle on tissue cells as the locus of respiration. Theoretical issues, including a variety of questions such as the role the tissues actually play in living organisms and the factors determining the rate of respiration, were also important. Bernard and Ludwig offered essentially theoretical arguments which they took to show the role of the blood in respiration. These were countered by Pflüger, who offered a convincing account of how the cells figure in respiration and how they are able to control the rate of respiration. In section 4 we explore how Pflüger was able to bring this controversy to an end and establish the tissue cells as the locus of respiratory phenomena.

Determining the locus of control is the first critical choice-point in the development of the mechanistic program. The cases we examine in Chapter 3 show some of the arguments that figure in controversies over this issue. As will be true of additional choice-points explored in subsequent chapters, one seldom has definitive evidence for the decision one makes; to some extent it will be made on the basis of other factors. Yet the choice is critical, for it determines the course of subsequent research. If one rejects some proposed locus of control, the task becomes one of identifying an appropriate alternative source of control—a challenge still confronting those who have stressed external factors in explaining behavior and evolution. If one accepts the identification of one locus, the next task is to seek the mechanism that operates within it and allows it to produce the function.

In Chapter 4 we explore what is frequently the first strategy pursued after a system has been identified—or accepted—as a locus of control. This is to identify a component within the system as itself responsible for the phenomenon, without yet inquiring how that component produces the effect. This, as we have said, we term simple or direct localization, because it involves localizing responsibility for an effect in a single constituent component of the system and then showing a direct link between the behavior of the system and that of the component. In developing a direct localizationist research program, what investigators look for are *correlations* between the performance of the system and the activities of one of its components. We consider two sets of cases in sections 2 and 3 of Chapter

4: Gall's localization of cognitive functions in the brain, and the attempt to localize responsibility for cellular respiration in different enzymes within cells.

Both examples of direct localization we consider are ones later researchers came to reject. This choice of cases is deliberate: As we said in Chapter 1, heuristics are strategies prone to failure. It is precisely where heuristics fail that their role in the development of science is clearest. Moreover, when decomposition and localization are deployed in the development of direct localizations, we observe the simplest and least demanding cognitive strategy. Given that scientists have finite problem-solving resources, it is not surprising that this strategy is where they often begin. Moreover, although it is probably seldom conceived of in this way by scientists themselves, the use of direct localization represents the pursuit of a strategy that, when it does fail, is likely to provide the most useful information about how the actual system is organized. This is a point we will develop more fully in Chapter 7.

The fact that direct localization was rejected in the cases we examine should not be taken to mean that it is never correct. Sometimes there *is* a single component responsible for the function of the system, and the direct localization program will be successful. Even when localization is successful, though, it is important to note that it does not itself produce a fully mechanistic explanation, because it does not explain *how* the function is performed by the component. At best it is a preliminary step, though one that is often taken in the research community and is important for the development of a field. The decision to use a different strategy would set an entirely different research agenda. Given a successful localization, perhaps confirmed by subsequent research, the next task is to explain how the localized function is realized. This requires a shift to a lower level. We consider such a case in Chapter 6. Research may also show that other components are involved and lead to a different kind of explanation, involving multistep models, at the same level. This is no longer simple localization. We consider such a case in Chapter 7. What is significant is how frequently investigators begin with a direct localizationist program, and what can lead to its abandonment or displacement.

In choosing whether to engage in such a program, researchers confront a second choice-point: Can one component account for the system's behavior? In answering this question affirmatively, investigators have implicitly affirmed that the system is decomposable. The mechanistic program depends on direct localization, proof of which will be given in Chapter 5 when we examine the arguments of the most radical critics of this approach. These critics claim that the correct explanation of a function requires that a system be nondecomposable, and they therefore reject decomposition and localization as strategies; this is an antimechanistic stance. Those objecting to these strategies offer a variety of positive proposals broadly classed as *dualist* in the case of psychological functions, or *vitalist* in the case of physiological functions.

Again, we look at two sets of issues. The first, discussed in section 2 of Chapter 5, involves the repudiation of Gall's phrenology by Flourens, with the support of Georges Cuvier. In part Flourens's attempt to refute Gall was experimentally grounded in lesion studies, in which Flourens emphasized what he referred to as the "unity of the nervous system." The second set of issues, discussed in section 3 of Chapter 5, focuses on vitalist opposition to mechanistic physiology in the first two-thirds of the nineteenth century. In particular we focus on Bichat's repudiation of Lavoisier's research, and on Schwann's and Pasteur's repudiations of the mechanistic program in fermentation. These attacks emphasize the central weakness of direct localization: its limitation to a correlational method and, consequently, to an impoverished empirical basis. The theories of Gall and Lavoisier *are* mechanistic, but they are not subject to systematic constraints from lower-level theories that could justify the decomposition into functions by demonstrating that there are lower-level components in which the functions could be localized. Indeed, Gall's and Lavoisier's models were developed largely without independent knowledge of component behavior—or even knowledge of the relevant components themselves. The decompositions into functions were a projection from the behavior of the system as a whole. This is mechanistic in inspiration but incomplete in its realization. As a result, the theories of Gall and Lavoisier are attacked by their opponents, such as Flourens and Pasteur, for being speculative and hypothetical.

These challenges to mechanism indicate the existence of an additional choice: whether to accept, even as a first approximation, the decomposability of the system. Researchers are sometimes led to give a negative answer at this choice-point by evidence that they take to show the failure of the direct localizationist program. Such individuals typically have independent grounds for thinking that one cannot develop an explanation by decomposing the system—a possibility we shall explore later. The negative decision, however, is a significant one, as it necessitates abandoning the mechanistic program for a different objective of research—a focus on *describing* the behavior of certain kinds of systems and delineating their properties, rather than *explaining* how such systems function. It is no accident that the major antimechanistic opposition to both Gall and Lavoisier, as well as to Geoffroy and Lamarck, came from within the French Academy. It was within the Academy, under the leadership of Cuvier, that the most strident opposition arose to Enlightenment materialism. The opposition to Gall and to Lavoisier can be seen as an extension of this antimaterialist and antimechanistic stance.

CHAPTER 3

Identifying the Locus of Control

1. INTRODUCTION: IDENTIFYING SYSTEM AND CONTEXT

Before developing a mechanistic explanation of a particular phenomenon, one must identify which system is responsible for producing that effect. Identifying a responsible system presupposes several critical decisions. The scientist must segment the system from its context and identify the relevant functions assigned to it. To substantiate the assignment of a function to a system, the scientist generally must offer theoretical or empirical arguments showing that the physically and functionally independent system identified has substantial internal control over the effect. This is what we describe as treating the system as the locus of control for a phenomenon. That some system or component of a system is a locus of control for a particular phenomenon does not entail that it is able to produce the effect entirely in isolation; causal control is contextual. Moreover, the systems dealt with in the life sciences are typically not closed, but open, systems; and they are not simply decomposable, but interactive. A locus of control for a given effect is a system or a component of a system that carries out the processes relevant to realize the effect. This is, in essence, what it is to be a machine. An example might help: An automobile does not produce motion on its own. It requires gasoline, a driver, a platoon of mechanics, large amounts of money, etc. Yet, it is *in* the automobile that the chemical energy is transformed into mechanical energy, and so the automobile is the appropriate locus of control for this effect.

In distinguishing a system as a locus of control we make the same assumption as when we explain the activity of a system in terms of the functions of its parts. We assume that nature as a whole is decomposable into units and that the system we are identifying is such a natural unit. In treating it as the locus of control, we assume that variations in the mechanism will be reliable indicators for variations in its behavior and will thereby explain them. Finding the right system is often difficult. It is equally difficult to find the right level of organization and the right boundaries to the system. On some occasions nature seems to divide naturally into systems. Generally this reflects what Wimsatt (1980b) calls "environmental grain" and is largely a result of perceptual and information-processing capacities we as researchers have developed and bring to the context. We readily discriminate organisms—at least organisms of a certain size—as entities, though a swarm of insects may be treated as an entity rather than an aggregate. A bat or a bird, by contrast, can recognize individual insects within a swarm. If we have the capacity to identify and isolate a system, this is usually because, for one reason or another, it is important for us to recognize or coordinate our lives in response to it, or at least to objects of the same scale.[1] Most animals are able to recognize individual conspecifics and members of a variety of other organisms in their environment. These natural capacities make the breaks in nature seem transparent, yet the "natural" breaks may be different from the apparent ones, and these are apparent only because they are important to us. The significant boundaries for scientific inquiry may be quite different than those that are important for other human activities; researchers must discover and learn to recognize these boundaries.[2] In most domains of inquiry, such recognition evolves with time and research as scientists develop conceptual frameworks to determine a particular way of decomposing nature into systems. Until such frameworks are developed, or when they are subsequently brought into question, decisions over how to divide nature into systems can be quite controversial.

The fact that disputes often arise over dividing nature into systems, and then over situating the locus of control for various phenomena in these systems, can be helpful in understanding theory development. Briefly examining these controversies will highlight the issues and the character of the decisions that are made when such conflicts are confronted. These issues and decisions are often concealed once the question of identifying a locus of control has been resolved. This is natural, as the decision virtually defines a research program or tradition, and a resolution will preclude alternative lines of inquiry. To explore these issues and decisions, we shall consider the two domains mentioned earlier in which there have been perennial conflicts over where to situate the locus of control: behavior and evolution.

In both cases, common sense offers a fairly natural division of nature into systems. Individual animals and biological species have long been identified as natural units, even if there has been considerable disagreement over their precise character; for example, the contemporary dispute over whether species are natural kinds or historical lineages, and traditional philosophical debates concerning identity across time, are both metaphysical disputes of this sort. The question here, however, is whether these commonsense boundaries mark off loci of control. This question is an *explanatory* one, as we want to explain such things as behavior and evolution. It is with respect to this explanatory question that controversies arise. Some researchers have argued that these systems are the appropriate locus and that what is outside these boundaries, the environment, can be treated largely as background or as secondary in significance. The consequence of this internal localization is an emphasis on research into the internal mechanics of these systems in order to explain them. Others argue against situating control within the systems themselves and contend that the important controlling variables are to be found outside of these systems. Those who deny that the system is the locus of control commonly argue for the opposite extreme and treat the system as relatively insignificant in producing the phenomenon in question. For advocates of internal localization, the system looms large and the context is of vanishing explanatory value. For their opponents, the context looms large and the system merely processes inputs.

Our purpose is to elucidate the kinds of considerations that are important in settling on a locus of control for a given phenomenon. We do not provide an exhaustive taxonomy. We offer examples only. The brief discussions in sections 2 and 3 of this chapter should illustrate the question and the principal approaches to answering it. In section 4 we will turn to a more extended example from physiology—the identification of the cell as the locus of biological respiration. In this case there was active controversy during the first three quarters of the nineteenth century until the case for the cell as the locus of control was finally established and researchers were able to direct their attention into the cell to explain how it accomplished the respiratory function. Here there was a resolution to the question of the locus of control, and the problems then were transformed, or, perhaps, they evolved. The mechanical question became the focus of investigation.

2. EXTERNAL CONTROL:
THE ENVIRONMENT AS A CONTROL

Controversies over the control of behavior and of evolution display many similarities. In both cases there are those who favor extreme environmentalist views, arguing that the mind/brain or the species were basically pliable entities shaped by their environment. Advocates of this view often focus on how adaptive and responsive the system is to external demands. If the system is extremely responsive to environmental variation, then, it is reasoned, differences must be environmentally induced and controlled. The net result of emphasizing the external factors is to reduce the importance of the system, treating it as responding to external factors, or shaped by them, but not itself an important element in accounting for the responses. A variety of arguments are generally offered as to why the system is unimportant and can be skipped over in the quest for an ultimate explanation.

Radical Behaviorism: Watson and Skinner

Behaviorism as a research program in psychology has many roots. Two are crucial: the *associationistic* program derived from Hume and Spencer in philosophy, and the *functionalist* psychology of William James, John Dewey, and James Rowland Angell, which dominated psychology at the turn of the century. Both downplay the importance of mental structures and construe mental activity as largely adaptive to external processes.[3] The associationist program is committed to the view that even the most complex structures constituting our knowledge of the world are constructed from sensory experiences, with simple general mechanisms emphasizing the association of ideas and experiences. The functionalism of James, Dewey, and Angell was born of Darwinism and emphasizes the adaptive role of cognitive processes:

> The functionalist psychologist . . . in his modern attire is interested not alone in the operations of mental process considered merely of and by and for itself, but also and more vigorously in mental activity as part of a larger stream of biological forces which are daily and hourly at work before our eyes and which are constitutive of the most important and most absorbing part of our world. (Angell 1907, p. 88)

By contrast with the *structuralist* views of Wilhelm Wundt and E. B. Titchener, classical functionalism concerns itself with the role of mental processes in regulating behavior, and downplays the importance of introspectionist taxonomies of mental acts and contents, such as those promoted by Franz Brentano. This role, in turn, is to be understood in terms of how behavior is modified and controlled in a natural setting. Functional psychology, Angell tells us, "portray[s] the typical *operations* of consciousness under actual life conditions, as over against the attempt to analyze and describe its elementary and complex *contents*" (ibid., p. 85). For classical functionalism the contribution of the environment in shaping our mental structures is fundamental.

In what is rightly regarded as the manifesto of radical behaviorism, John B. Watson (1913) initiated a movement that was to dominate psychology for roughly the next fifty years. With his functionalist progenitors, Watson emphasized adjustment to the environment as the primal fact of psychology.[4] The internal causes, whether physiological or mental, were unimportant for the purposes of a scientific psychology. Psychology, so understood, is concerned with the ability to modify and control behavior in the face of environmental demands. Behaviorism, Watson declares, "is the only consistent and logical functionalism" (ibid., p. 514). Behavioral changes are under the control of external stimuli: "stimuli lead the organisms to make the response" (ibid.). The only behavioral changes with psychological significance are *responses* to environmental changes; moreover, our behavioral capacities are the result of environmentally induced modifications in learning, and it is in terms of the simple learning mechanisms that complex capacities must be understood.

The emphasis on environmental control, which is characteristic of behaviorism, in no way implies that internal mechanisms are unimportant. Behaviorists only deny internal mechanisms' significance as primary controlling variables for the purposes of psychology. Watson says quite clearly:

> Much of our structure laid down in heredity would never come to light, would never show in function, unless the organism were put in a certain environment, subjected to certain stimuli and forced to undergo training. Our heredity structure lies ready to be shaped in a thousand different ways—the same structure—depending on the way in which the child is brought up. (1924, p. 97)

The pivotal commitment of behaviorism is the view that our complex behaviors can be neither explained nor understood in terms of internal mechanisms. Any given behavior will be a function, inter alia, of the environment together with the capacities we have developed; these capacities, in turn, will be a function of our developmental history together with innate mechanisms. Yet these innate mechanisms are so simple and general that they radically underdetermine the result. The hereditary structure, as Watson says, "lies ready to be shaped in a thousand different ways." As a result, the specific responses or patterns of responses can be explained, if at all, only in terms of environmental variables.

The rejection of *instincts* as determinants of behavior—Watson devoted two substantial chapters to the topic in *Behaviorism* (1924)—is not a denial of the existence of instincts as much as a denial that they are sufficient to explain particular behavioral results. "The central principle of behaviorism," as Watson tells us, is that "all complex behavior is a growth or development out of simple processes" (ibid., p. 137). This is a legacy of associationism. Its heir, in turn, is the learning theory of Tolman and Skinner. So conceived, we are born with simple capacities to respond in determinate ways to determinate stimuli.[5] Over time, these capacities develop into increasingly complex abilities under the influence of simple learning mechanisms. Watson emphasized respondent conditioning, allowing substitution of both stimuli and responses. Skinner (1938) introduced operant conditioning, with the express intention of explaining behavior for which there is no evident elicitation by a stimulus.

The commitments of behaviorism are, thus, varied. They include a modified associationism with simple and general mechanisms of learning, but reconstituted to apply to behavior rather than to ideas; the rejection of instinct as a significant explanatory concept for psychology; and the commitment to learning theory, with learning mechanisms common to humans and other animals (cf. Watson 1913, p. 507). Together, these commitments constitute a single fabric emphasizing environmental control of behavior. Each requires the others.

Parallel commitments are equally evident in B. F. Skinner's treatment of operant conditioning in *The Behavior of Organisms* (1938). Two of the central "dynamic laws" governing *operant behavior* (that is, behavior that appears to be spontaneous) pertain to conditioning and extinction. In Skinner's own terms,

> If the occurrence of an operant is followed by presentation of a reinforcing stimulus, the strength [of the operant behavior] is increased.
>
> If the occurrence of an operant already strengthened through conditioning is not followed by the reinforcing stimulus, the strength [of the operant behavior] is decreased. (Ibid., p. 21)

These are the "laws" of operant learning in a qualitative form. They describe changes in behavioral tendencies solely as a function of environmental variables. Reinforcement will increase the "strength," understood in terms of persistence, of some behavior; withdrawal will decrease it.

In the cases of both Watson and Skinner, behaviorism emphasizes environmental control. Innate mechanisms are incapable of explaining the adaptive responses of any organisms, or the corresponding range of their behavior. Indeed, variations in whatever innate mechanisms there are should create no qualitative differences: all differences are differences of degree only. Internal mechanisms that mediate the control of learned behavior must themselves be explained; ultimately, this requires an appeal to environmental conditioning or to selection. This emphasis on external control thus depends on the commitment to simple and general mechanisms of learning and, correlatively, rejects the mental system as, in any interesting sense, a locus of control for behavior.

Natural Selection and Adaptation

In a parallel fashion, the Darwinian emphasis on natural selection and adaptation embodies an externalist approach. Darwin was faced with two related problems in *On the Origin of Species* (1859). On the one hand he was confronted with the task of justifying the claim that evolution—or, as he preferred to describe it, the "transmutation of species"—occurred. On the other hand he proposed to defend a particular mechanism for the evolutionary process. That mechanism, in turn, was simultaneously to explain transmutation and the "perfection of structure," which we have since come to think of as adaptation. As Darwin himself wrote in the first edition,

In considering the Origin of Species, it is quite conceivable that a naturalist, reflecting on the mutual affinities of organic beings, on their embryological relations, their geographical distribution, geological succession, and other such facts, might come to the conclusion that each species had not been independently created, but had descended, like varieties, from other species. Nevertheless, such a conclusion, even if well founded, would be unsatisfactory, until it could be shown how the innumerable species inhabiting this world would have been modified, so as to acquire that perfection of structure and coadaptation which most justly excites our admiration. (Ibid., p. 3)

The central problem for Darwin was thus one that was also common to natural theologians in the tradition of Paley and the Bridgewater treatises. Like them, Darwin insisted on taking seriously the adaptation of organisms to their environment. He differed in *how* to explain this adaptation, but the emphasis on adaptation as well as transmutation led him to locate the control of evolution outside the species and in the environment.

Darwin developed the case for natural selection as the prime mover of evolutionary change in the early chapters of the *Origin*. The general argument is straightforward and simple: Organisms exhibit a remarkable degree of adaptation to their environment, as well as to other species, which are factors in that environment. This adaptation, or coadaptation, is explicable in terms of natural selection if it is understood to operate on individuals over large expanses of time. Natural selection, or the differential survival of individuals on the basis of variations in fitness, is capable of explaining the kind of finely tuned adaptation Darwin saw as so central to the natural order, provided that variations are small, many, and heritable.

Darwin himself adopted an eclectic view and incorporated a variety of secondary mechanisms besides natural selection, among them the environmental induction and the inheritance of adaptive variation which is commonly, if misleadingly, referred to as the "inheritance of acquired characteristics."[6] The same commitment to external control can be seen in Darwin's treatment of the variation on which natural selection acts. He consistently maintained, first, that environments will induce variation in organisms when the species is marginally adapted for those environments; and, second, that when variation is induced by the environment, all organisms in the species encountering that environment will vary in the same manner.[7] In discussing the case of an environment undergoing continual change, Darwin claims that individuals of a given species within that environment will tend to undergo similar changes:

Changes in the conditions of life give a tendency to increased variability; and in the foregoing cases the conditions have changed, and this would manifestly be favourable to natural selection, by affording a better chance of the occurrence of profitable variations. Unless such occur, natural selection can do nothing. (1872, pp. 75–76)

The induction of variation in marginal environments had the advantage of increasing the amount of variation in just those places in which it would do the most good, and thus could increase the rate of evolutionary change as well as the degree to which it could maintain adaptation.

The problem of explaining adaptation, inherited from Paley, thus lies at the center of Darwin's case for natural selection as the mechanism of evolution. Since what is to be explained is adaptation to an environment external to the organism, the control for that adaptation must either lie in that environment or be due to some agent that can anticipate and guide the change. Darwin took the former option without hesitation.

In more recent Neo-Darwinian work there has been a parallel emphasis. The view that genetic variation enters the species through chance-like mutations, is passed on from one generation to the next, and is selected as organisms compete for survival and reproduction became the orthodox account of evolution with the development of population genetics in the 1920s and 1930s. The working hypothesis is the assumption that a structure or behavior is an *adaptation,* something that facilitated survival and was promoted in the population precisely because it did so (cf. Brandon 1978). In theory it is allowed on all sides that not all traits are adaptations; some, at least, will have other sources. In practice, however, many biologists resort to nonselectionist explanations only as a last resort. Without serious consideration of what would otherwise be expected—in short, without attention to qualitative "base rates"—nonselectionist mechanisms are set aside. Questions about how organisms develop, or about the causal processes active in individual organisms, are set aside as if they were of no interest and did not have an important influence on the overall course of evolution. Ernst Mayr (1961), for example, distinguishes between proximate and ultimate causes of evolutionary change, and, while granting that there may be a variety of proximate and local factors that determine the character of members of a species, he argues that these are not the true determining variables. For that, in his view, we must turn to natural selection.

The point may be underscored by turning to one of the more widely accepted cases in recent theorizing, R. A. Fisher's (1930) explanation of the fifty-fifty sex ratio present in most sexually reproducing organisms. Fisher assumes that either parent can induce a bias in the sex ratio of immediate offspring, skewing the frequency of one sex in either direction from the norm, and further assumes that the reproductive investment involved in bringing a male to maturity is different from that necessary to bring a female to reproductive maturity. He then argues that the optimal strategy would be to equalize total expenditure of reproductive resources on males with the total expenditure on females. (The details of his argument need not concern us here.[8]) We may grant that, under the stated assumptions, a fifty-fifty ratio will be most common. The question is whether this is an adaptation. There are obvious alternatives to the selectionist explanation that are at least as plausible and are not even considered as candidates (cf. Gould and Lewontin 1979; Lewontin 1978). For example, a fifty-fifty ratio will be the natural consequence of gametogenesis—the production of sexual gametes—because this involves the partitioning of genetic material into two reproductive cells. It is in fact possible that a fifty-fifty sex ratio is not an adaptation at all, but the natural, if not inevitable, consequence of the reproductive machinery in bisexual organisms. The assumption that it is an adaptation bespeaks the commitment to natural selection and to external control.

3. INTERNAL CONTROL: THE SYSTEM AS A CONTROL

Each of these externalist programs has had opponents arguing in favor of internal control. As would be expected, the contrast between contenders is multidimensional. One common thrust to internalist arguments is that the system in question is not as flexible and adaptive as externalist theories require: Response is limited and structured in the face of large variation. The phenomenon that requires explanation is the limitation on the range of response. Limited responsiveness in the face of wide environmental variation is taken as indicative of internal control, and the solution is to search for specialized and complex internal mechanisms. The system makes its own contribution and influences what happens to it.

The Mentalist Program in Linguistics

Just as *behaviorism* has its roots in the associationist program, which limited internal activity to simple and general procedures for associating ideas, the *mentalist* program traces its roots in part to a rationalist account of knowledge, with a corresponding emphasis on the contribution of cognition to the constitution of the world. Just as behaviorism stresses the external control of behavior, mentalism emphasizes the indispensability of psychological determinants in explaining behavior, many of which are argued to be innate, not learned. This means, in principle, that there is an indeterminacy of behavior when only external parameters are included, and, in practice, that there is an inability to predict or explain important dimensions of behavior in environmental terms alone. A theory emphasizing external control will be correspondingly incomplete, and it will be necessary to adopt an approach that makes the mental system itself the locus of control.

The history of psychology provides ample evidence to support the mentalist contention. Learning theorists, including behaviorists, were committed to the view that animal learning and human learning differed only in degree (see Skinner 1938, pp. 441–42). Yet the sort of simple mechanisms that were constitutive of the various models failed to generalize to more complex behaviors, including many exhibited by humans. Indeed, once removed from the limited setting of psychological laboratories, it became difficult to isolate or describe the variables supposedly controlling behavior. Skinner's own vacillation on the defining characteristics of behavior (see Chomsky 1959; Scriven 1956) in more popular works (for example, *Science and Human Behavior*, 1953), in contrast to more technical works such as *Verbal Behavior* (1957), is explicable as an immediate consequence of the failure to isolate classes of behavior and environmental variables with suitable functional relations. The extension of conditioning models to complex behavior was a failure. The only alternative was to treat the cognitive system seriously as a locus of control.

By the middle of the twentieth century, the shortcomings of behaviorism had become all too apparent and were crystalized by Noam Chomsky in his argument from what he called the "poverty of the stimulus" (cf. Chomsky 1980, pp. 35ff.). In its simplest form, this argument presses that the specific characteristics of human language are underdetermined by, and inexplicable in terms of, environmental variations alone:

> Gross observations suffice to establish some qualitative conclusions. Thus, it is clear that the language each person acquires is a rich and complex construction hopelessly underdetermined by the fragmentary evidence available. (Chomsky 1975, p. 10)

Children have limited information given to them about the language they are learning. What information they do get, moreover, is fragmentary and disconnected. Yet in a matter of months they develop an elaborate and detailed understanding of their native language. The moral Chomsky draws is straightforward:

> The essential weakness in the structuralist and behaviorist approaches to these topics is the faith in the shallowness of explanations, the belief that the mind must be simpler in its structure than any known physical organ and that the most primitive of assumptions must be adequate to explain whatever phenomena can be observed. (1968, pp. 25–26)

If the environment leaves the particular details of language "hopelessly underdetermined," then the explanation of these details must inevitably lie in the structure of the mind. As a consequence, Chomsky is committed to embracing innate, internal mechanisms underwriting language learning. This is a rationalist program with an internal locus of control; it does not deny the significance of the environment any more than the behaviorist denies the importance of internal mechanisms. We do, after all, *learn* the language our parents speak—the ability to speak as they do does not lie in our genes. Chomsky's point, rather, is that the specific details concerning the what and how of learning can be explained only by placing the explanatory burden on internal mechanisms.

The case for internal control just sketched was given a more precise formulation by Chomsky in *Syntactic Structures* (1957) and *Aspects of a Theory of Syntax* (1965). The argument Chomsky uses is a linguistic variant on one that has driven mentalism from Descartes to Kant: If we think of a grammar for a language as a set of rules, then, at a minimum, an adequate grammar must be capable of producing all (and only) the strings that are acceptable in that language. Primary-language learning will then involve inducing a grammar from some finite and limited set of sentences presented to the child. Since any finite set of sentences can be generated by an infinite number of grammars that are, nonetheless, formally nonequivalent, the task of language learning requires selecting the correct grammar from among the set of abstractly possible grammars. The only alternative is to allow that we incorporate some mechanism that enables us to limit the number of candidate grammars. That is, there must be some innate structure to guide and inform language learning in the child. Moreover, natural languages are sufficiently complex that the grammar we end up with cannot be a simple algorithm.[9] If grammar is not a simple, general algorithm, then, because it will be largely accidental which heuristic we employ, the mechanism for language will likely be species-specific. The nativism that plays such a central role in Chomsky's thought is driven by the need to accept just this sort of complex, innate, internal structure as the foundation for language learning. Thus, in contrast with behaviorism, Chomsky argues for the mental system as the locus of control with respect to linguistic behavior.

Developmental Control of Evolution

Just as the nativism of Chomsky serves as an alternative to the environmental emphasis of behaviorism, so there is an analogous alternative to the environmentalist orientation of Darwinists and Neo-Darwinists. Its roots lie in *preformationist* theories of development, which were prevalent toward the end of the eighteenth century and early in the nineteenth. Whereas *epigenesists* maintain that form emerged gradually, in the developmental process, preformationists maintain that the fully differentiated animal form is predetermined at the earliest stages of development. Epi-

genesists portray development as a transition from an organism undifferentiated with respect to form to a fully differentiated organism—from an
unorganized entity to an *organism* proper—under the influence of environmental forces. Such forces literally mold the organism into the form it
assumes. Preformationists, by contrast, portray development as an unfolding of what is latent in the individual—from one organized form to
another—largely independent of the environment. As Jane Maienschein
explains the preformationist view, "Development remains strictly internally determined and not subject to external or environmental conditions
in any significant way" (1986, p. 4).[10]

Under the influence of theories that viewed individual development as
recapitulating the evolution of the species, Darwinism became transformed into an analogue of preformationist development (cf. Gould 1977;
Maienschein 1978); evolution became the unfolding of developmental patterns that were reflected and revealed in ontogeny. As Haeckel writes,
"Phylogeny is the mechanical cause of ontogeny" (1874, p. 5). Just as development reveals increasingly complicated forms of organization, so, too,
does evolution. Evolutionary change involves, according to this recapitulationist view, the successive addition of stages to ontogenetic patterns
already established, and then the subsequent attenuation of these stages
(cf. Gould 1977, ch. 4). Terminal addition and acceleration of development then form the basis for the next round of evolutionary modification.

In the earliest Neo-Lamarckian writings in the United States, the recapitulationist view takes a developmentalist turn under the leadership of
Edward Drinker Cope and Alpheus Hyatt.[11] The tradition became one of
the first systematic challenges to Darwinism, and it persisted even into
the twentieth century. In its earliest years the view was Lamarckian insofar as it was committed to an orthogenetic model of evolutionary change:

> Cope and Hyatt began from Agassiz's view that the growth of the individual
> offers a model for the history of life on earth. They accepted that the pattern of
> development revealed by a group's fossil record is recapitulated in the growth
> of the modern embryo. Evolution proceeds step by step through the addition of
> stages to individual growth, and the pattern of development is essentially pre-
> determined and regular. (Bowler 1988, pp. 99–100)

The major point of contention between Neo-Lamarckians and Darwinians
was the relative importance of natural selection in the evolutionary process, and not the importance of evolution in the origin of species. While
Darwin and his closer followers (such as Asa Gray and Thomas Huxley)
claim natural selection as "the main but not exclusive means of modification" (Darwin 1859, p. 6), Cope and Hyatt maintain that the crucial feature in explaining evolution is the *origin* of favorable variations rather
than their preservation. As Cope writes, "Nothing ever originated by nat-

ural selection, . . . [and] important though it is, [it] is but half the question, and indeed the lesser half" (1887, p. 16). Adaptation to external circumstance is relegated to a secondary role.

According to Cope and Hyatt's view, recapitulation results from developmental mechanisms that would explain the pattern to be found in evolutionary change.[12] Evolutionary progress is marked by the addition of developmental characters that originated in previous phylogenetic stages. Because what develops before maturity can be inherited, an acceleration in development will result in the hereditary acquisition of these new characters. Ontogenetic stages would thus recapitulate phylogenetic changes. Acceleration of development, then, incorporates newly acquired traits and allows them to be inherited. The loss of characters in retrogressive change, in an analogous way, results from a retardation of development, with a resulting reduction of organic structures and complexity.

For our purposes here, the details of the Neo-Lamarckian account, as well as its motivation and development, are less important than the central explanatory strategy adopted by these opponents of Darwinism. Neo-Lamarckians emphasize internal developmental influences on evolutionary change as more important than the environmental influences on development of, at least, generic characters. The explanation for major changes is thus internalist, reflecting the view that evolutionary changes are guided largely by developmental rather than environmental processes. Thus, in contrast with more orthodox Darwinians, Neo-Lamarckians treat the species as a locus of control for evolutionary change. Through the early twentieth century there was a decreasing emphasis on the importance of development for evolution. The environment was on the upswing as an explanatory variable. The issues, however, did not die, and have recently come again to the fore.[13]

4. FIXING ON A LOCUS OF CONTROL: THE CELL IN RESPIRATION

We turn now to a case in which a long-lasting controversy over locus of control was eventually settled. It involves the biological process of respiration; that is, the process employing oxygen in reactions with foodstuff.[14] This controversy was ultimately resolved in favor of structural units within organisms—the cells found in biological tissues. Knowing the physical identity of the cell, however, could not settle that it was the locus of control for the function, or what its function was within the organism. In fact, by the time the controversy over the localization of respiration was sharply focused in the late nineteenth century, cells had been distinguished as both structural and functional units, largely through the investigations of Theodor Schwann (see Bechtel 1984a). The functional considerations most central to Schwann's (1839) account are those concerning growth. He argues that the structures in different tissues of animal bodies are all the same kind of physiological units—that is, cells—on the basis of the fact that they form in the same way. Schwann also argues, though, that basic metabolic processes, which would include respiration, are performed within the individual cell. He contends that because these functions have to be performed *within* the cell in single-celled organisms, and there is no reason for them to be performed within cells in one context and outside of cells in another, the cell has to be the locus of metabolic functions.

Neither Schwann's arguments nor the acceptance of cells as basic structural units of living organisms, however, were decisive on the issue of the locus of control for respiration. Before turning to the later stages of the controversy and its eventual resolution, though, it will be useful to consider its origins. They lie with Lavoisier.[15] After advancing the oxygen theory, according to which combustion (then reconstrued as oxidation) involved the combination of oxygen with hydrogen or carbon, rather than the release of phlogiston, Lavoisier joined with LaPlace in a study of animal respiration. They argued that animal respiration was also a form of oxidation by demonstrating that the heat output of animals (measured by their ability to melt ice) was comparable to that of ordinary combustion:

> By comparing the heat evolved by the combustion of carbon with the quantity of fixed air which is formed in this combustion, we have the heat developed in the formation of a given quantity of fixed air; if we determine next the quantity of fixed air which an animal produces during a given time, we shall have the heat which results from the effect of respiration upon air; it then only remains to compare this heat with that which sustains the animal heat and which is estimated by the quantity of ice which it melts within our machines; and if, as we have found by previous experiments, these two quantities of heat are approximately the same, we can conclude directly and without hypothesis that the 'conservation' of animal heat is due, at least in its major part, to this change of the pure air into fixed air. (Lavoisier and LaPlace 1780/1862, p. 332)

Lavoisier and LaPlace concluded that respiration was "slow combustion," and they proposed that this combustion occurred in the bronchi of the lungs:

> Respiration is therefore a combustion, very slow it is true, but otherwise perfectly similar to that of charcoal; it occurs in the interior of the lungs, without producing perceptible light, because the liberated matter of fire is immediately absorbed by the humidity of these organs. (Lavoisier and LaPlace, 1780/1862, p. 331)

The proposal that combustion occurred in the lungs was not accepted by all. La Grange, for example, contended that the amount of heat produced by having all oxidation of foodstuff occur in the lungs would have destroyed the organ:

> If all the heat which is distributed in the economy was set free in the lungs, the temperature of the lungs would necessarily be raised so much that one would have reason to fear that it would be destroyed. (quoted in Foster 1901/1970, p. 252)

La Grange's objections, however, were not viewed as decisive. In fact, Berthelot (1889) much later challenged La Grange's calculations, arguing that rapid circulation of blood and air in the lungs would mean a net rise of less than one degree. But those who took objections such as La Grange's seriously sought a different locus for respiration, and most focused on the blood.

The debate over the site of respiration was not just a theoretical issue; it was also addressed at the experimental level. One important experimental strategy (which is an analogue to the analytic strategy introduced in Chapter 2 and which will become especially important in subsequent chapters) was to attempt to show that respiration can occur in particular sites even in isolation from others. It was this line of experimentation that began to point to tissues as a third possible site of respiration. Vauquelin (1792) confirmed respiration in insects using the same model as Lavoisier. Because insects lack lungs, Vauquelin reasoned, there must be an alternative locus for respiration—which he argued was the stomach. Spallanzani, whose work was posthumously published by Senebier (1803, 1807), performed extensive experiments on respiration in molluscs and crustacea. He showed that formation of carbon dioxide could occur in these organisms even when they were deprived of fresh oxygen. He also demonstrated that different organs (lungs, brain, flesh, liver, and skin) absorb oxygen and give off carbonic acid (CO_2). Clearly, this told against the lungs as the sole or primary locus of respiration. Spallanzani also argued theoretically against respiration in the blood on the grounds that blood was incapable of carrying out the reaction.

> The blood is not of all the animal parts the one most suited to the destruction of oxygen gas, although at first, judging from what has been written in the subject relative to the decomposition of air, I believed that it exceeded all the others. Blood, arterial and venous, from warm and cold-blooded animals, has been tested, and I have never had any variation in the results. (1803, p. 86)

Altogether three candidate sites for respiration were available, and the task investigators faced was to find evidence that could rule out one or more of the loci or, conversely, could show that one of the loci did constitute the major site of respiration. Two major considerations figured in the debates during the first half of the nineteenth century: whether there were sufficient thermal differences between the lungs and rest of the organism to allow for respiration to be localized in the lungs, and whether oxygen could leave the lungs to travel through the blood and potentially into the tissues. Without significant thermal differences, the lungs could not be the sole locus of respiration; without a mechanism of transport, they had to be.

The best evidence available at the time on these two questions pointed in different directions. J. Davy's (1815) failure to find any significant thermal differences between arterial and venous blood stood against the suggestion that respiration occurred in the lungs. On the other hand there were problems explaining how oxygen might get from the lungs to the blood. Ellis (1807) examined the possible modes of transfer and found them all wanting. He noted that the surface of the lungs was covered with absorbent vessels, but contended "that the fineness of these vessels, the mucus perpetually smearing the surface of the cells, the elastic nature of the air itself, and its repulsion by water, so that it neither penetrates moist paper, cloth nor skin—all demonstrate that no air by this route gets in the blood" (1807, pp. 117–18). Ellis also considered the possibility that the transfer of oxygen was effected by the power of chemical affinity, but he rejected the existence of such an affinity, remarking that "even granting to the blood this power of attracting air, or its oxygenous portion, it is not easy to conceive why it should so readily lose it and again give out this air in the form of carbonic acid" (ibid., p. 123).

In the 1820s and 1830s the problem of oxygen transportation was resolved. Dutrochet separated two fluids of different densities with a membrane and showed that a bidirectional flow occurred. Graham demonstrated the same result with gases. Using membranes from fowl, Faust and Mitchell showed that oxygen and carbon dioxide could readily pass between the lungs and the bloodstream, thus overcoming the objection that had been raised against earlier proposals. The mechanism for transferring oxygen from the lungs to the blood, combined with the thermal evidence, eventually led to the rejection of the lungs as a candidate site for respiration. This shift, however, was not immediate: Magendie and Bernard were both still actively investigating the lungs as a possible site of respiration in the mid-1800s. This investigation, in which Bernard (1856) followed up on previous work by Magnus (1837), offered additional thermal evidence that arterial blood was not warmer than venous blood and seems to have finally established that the lungs did not control respiration. One alternative was eliminated.

Of the two remaining options—the blood and the tissues—the blood was favored by most researchers as controlling respiration, as theories then in vogue portrayed the blood as the center of physiological activity. Moving to the tissues as a site of oxidation meant reconsidering the tissues—generally taken to be passive—as active sites for metabolic function. Several proposals emerged in the 1840s supporting the localization of metabolic processes in the tissues. Liebig, in formulating his general account of metabolic processes, assigned important functions to the tissues. Moreover, Helmholtz (1847) showed that the contraction of muscles produced heat; that, in turn, suggested that metabolic changes were occurring in the muscles. However, the question of *where* metabolic processes occurred turned out not to be fundamental to determining the site of respiration. Researchers considered the possibility that respiration might occur away from the site where metabolic reactions released energy. The primary focus in the arguments over the site of respiration turned on how and where gases could be transported, not on where the metabolic processes associated with respiration occurred (Culotta 1970a, b). Thus, in the 1850s and 1860s a variety of investigators pursued the issue of the locus of respiration by addressing the question whether gases, and particularly oxygen, could move from the blood into the tissue cells.

Georg Liebig (1850), Justus Liebig's son, focused on the movement of gases. He studied respiratory processes in muscles under a variety of conditions—for example, when saturated with water and when kept in a variety of atmospheric conditions—and concluded that muscles do take up oxygen and release carbon dioxide. He argued that the blood acted solely as a means of transport to the tissues, with the carbon dioxide forming within the tissues and passing into the capillaries. The younger Liebig's experiments showed that the gases required for respiration could reach the tissue cells and were therefore a genuine candidate for the locus of control for respiration. Experiments, however, failed to show that they were the actual site of respiration in living organisms. Two major investigators, Claude Bernard and Carl Ludwig, continued to oppose the tissue cells as the site—or, at any rate, as the major site—of respiration. While they based their arguments in very different ways on considerations about what actually controlled the rate of oxidation, both concluded that the blood was the principal locus of respiration.

Bernard was preoccupied with how the blood came to change color; he attributed it to respiration.[16] Bernard's conclusion was that the darkening of the blood resulted from the increase of carbon dioxide resulting from a chemical reaction occurring in the blood itself. He measured tissue respiration in vitro and compared the carbon dioxide production of liver, kidney, muscle, and brain (1859). Bernard, however, did *not* interpret the results showing respiration here as showing that actual respiration was performed by the tissues. He argued that Liebig's work was inconclusive because it did not rule out the possibility that blood corpuscles stored in the tissues accomplished the oxidation.[17] Moreover, he argued that only liquids were transferred between the tissues and the blood. His inability to liberate oxygen from red blood cells indicated to him that oxygen could not be passed from the blood to the tissue cells:

> If it is true, and we are much inclined to admit it is so, that the venous blood owes its black coloration to carbonic acid, we must recognize that the modification by which its oxygen could be transformed into carbonic acid, can be brought about directly in it and not directly by immediate contact with tissues. (1859, p. 339)

Bernard proposed that the tissues excreted a liquid containing carbon and hydrogen, rather than carbon dioxide, which was then oxidized by the red blood cells. He concluded:

> It is infinitely probable that the carbonic acid of venous blood results from an oxidation which is brought about within the red blood corpuscle itself. When the blood traverses the capillaries, there will be between it and the tissues not an exchange of gases but perhaps one of liquids. Following the new conditions which such an exchange would create, the oxygen of the red blood corpuscle would be partly used for the oxidation of the carbon of the corpuscle itself. (Ibid., p. 342).

Bernard did not regard the question of the site of respiration as an isolated empirical issue; rather, it fit into his developing mechanistic conception of living organisms as able to regulate their own activities without requiring the agency of a vital force. Bernard held that each of the components constitutive of an organism was integrated with other components and was regulated by what happened in them. The result of this interaction was the maintenance of what he spoke of as the "constancy of the internal environment." Given this view, he argued that the locus of respiration—that is, the center in which the crucial processes of respiration transpired—had to be distinct from whatever controlled the rate of respiration, since regulation required one component to act on others. Bernard took the nervous system in particular to determine the rate of respiration, and he attempted to demonstrate this through experiments in which he cut the chorda tympani and sympathetic nerve. He interpreted the results as showing that the respiratory system regulated the rate of respiration by limiting how much hydrogen and carbon the cells excreted into the blood. Reactions with oxygen, he claimed, therefore occurred in the blood. If, on the other hand, the reaction with oxygen occurred within the cell, there would be no regulation of the reaction. Once oxygen entered the cell, there would be nothing to prevent the oxidation of all the carbon and hydrogen available. So, to enable processes within the tissue cells to regulate respiration, it was necessary that the reaction mechanism itself be situated outside the tissues and, hence, in the blood.

Ludwig also opposed tissue respiration—though less adamantly—but his reasons for being reluctant to accept it were quite different. In contrast with Bernard, Ludwig argued that the blood controlled the respiratory process. His arguments (Ludwig and Schmidt 1869) were grounded in his attempts to develop a mechanical model of the exchange of respiratory gases.[18] He proposed that pressure gradients governed the movement of gases between different media, and he thought studies of gas concentrations would settle the issue of where respiration took place: wherever concentrations of carbon dioxide were highest would be the site of respiration, with concentrations decreasing as one moved away from this site. Ludwig, however, was unable to produce sufficiently precise measurements to settle the issue. This was partly because of empirical difficulties in comparing pressures of gases in tissues and in blood, and partly because of problems in accounting for the possibility that some of the gas might be held in a chemically bound state.

Although his primary approach did not yield definitive data, Ludwig came up with an alternative strategy for determining what factors regulated respiration. He developed improved experimental techniques for perfusing organs with defibrinated blood whose gas content could be carefully measured. He noted a correlation between the rate of blood flow, the amount of oxygen consumed, and the amount of oxygen reduced to carbon dioxide. On this basis he claimed that the amount of oxygen in the blood controlled the rate of respiration: "In these numbers the law once more declares itself, that oxygen consumption increases with the velocity of the stream" (Ludwig and Schmidt 1869, p. 38).

Having located the *control* of the respiratory process in the blood, Ludwig went on to argue that the process itself was also likely to occur there. He asserted, in part, that given the rate with which the blood passed through tissues, there was not time for the oxygen to pass from the capillary into the tissue: "If one realizes the time over which the oxygen disappears, it seems scarcely adequate to effect, by way of diffusion, the copious exit of oxygen from the disks through the vessel wall" (ibid., p. 36).[19]

The contrast between Ludwig and Bernard is interesting. While both argued for the blood as the site of respiration, their arguments were incompatible. Both appealed to the factors they took to regulate respiration; however, for Bernard the locus of the process could not be the same as the factor regulating the process, whereas for Ludwig they were the same. Bernard saw regulation as arising from interaction between a regulating entity and that which is regulated, a perspective not shared by Ludwig.

The major proponent of intracellular oxidation was Eduard Pflüger (1872, 1875). Pflüger's task was twofold. He had to show both that respiration did not occur in blood itself and that it could occur in tissues. In pressing that respiration was not carried out in the blood, Pflüger produced evidence that the blood—or the availablity of oxygen in the blood—does not control the rate of respiration. He performed experiments that purported to show that animals were not sensitive to large variations in the availability of oxygen, thus demonstrating that gas was not exchanged in accordance with differences in pressure. He also exposed asphyxiated blood to oxygen, reasoning that if blood contains readily oxidizable material, it should be plentiful in asphyxiated blood. He took the fact that asphyxiated blood did not produce oxidation upon exposure to oxygen as evidence that the blood *in* organisms would not either.

Pflüger went on to contend that cells and tissues do regulate the rate of respiration. In order to argue this it was necessary to counter Ludwig's claim that time constraints would prevent sufficient transfer of oxygen from the blood into the tissues. Pflüger calculated the pressures of oxygen and carbon dioxide at various points in the body and argued that, especially given the enormous capillary surface, oxygen and carbon dioxide could readily diffuse from blood to tissue or vice versa. Pflüger derived major support for his claim that tissues control the rate of respiration from work on insects showing that tracheal tubes filled with air penetrated directly into the tissues and cells and supplied them with oxygen. If blood was not necessary for respiration there, he argued, it must serve no more than a transport function. He concluded:

> I wish to state this once and for all that herein lies the real secret of the regulation of oxygen consumption throughout the entire organism: that the cell alone determines it, not the oxygen content of the blood, not the tension of the aortic system, not the velocity of the blood stream, not the mode of cardiac output and not the method of breathing. (1872, p. 52)

We could hardly hope for a stronger expression of localization.

Oertmann helped further enforce the view that tissues, not blood, were the determinants of the rate of respiration. He replaced the blood of a number of frogs with saline solution and pure oxygen. Their respiratory activity was unaffected for ten to twenty hours, and the frogs finally died after one to three days. Oertmann concluded, with the frogs dissenting, that "the oxidation processes of the frog undergo no change following the removal of its blood. The bloodless frog has the same metabolism as the frog with blood. The site of the oxidation processes is therefore the tissue, not the blood" (1877, p. 395).

Pflüger's work provided convincing evidence for the claim that tissue cells were the locus of respiration: he effectively removed the evidence that suggested that the rate of blood flow was the critical controlling variable in respiration, and his work on insects demonstrated that tissues were able to carry out respiratory functions without blood as an intermediary. The evidence that tissues can respire, combined with an account of how oxygen could diffuse into tissues and evidence that tissues actually controlled the rate of respiration, established tissue cells as the locus of respiration.

Pflüger recognized that ultimately one needed to explain *how* the cell carried out respiration and was able to regulate the process. It was not satisfactory simply to stop after identifying the cell as the locus. His ideas on this issue, however, were repudiated as speculative (see Glas 1979, pp. 85-87, for further discussion). Even so, he saw the importance of recognizing that the next step was to discover an internal mechanism capable of performing the task assigned to the system.

5. CONCLUSION: LOCALIZATION OF FUNCTION

Defining and isolating a locus of control is one of the first steps in a mechanistic understanding of the behavior of complex systems. The initial task in identifying a locus of control includes segmenting a system from its environment and showing that it is capable of performing the activity assigned to it. Deciding whether a system constitutes a locus of control is thus the first choice-point on the path to developing a mechanistic explanation. If one successfully differentiates a system as a locus of control, the next task is to determine how the system performs the required functions. If the system is not the locus of control, the task is once again one of segmenting an appropriate system that might be the locus of control for the phenomenon under investigation. This process may be represented as in Figure 3.1.

Segmentation of a unit generally requires collateral theories that refer to the system or indicate its structure. Sometimes the theories leading researchers to segment nature as they do and distinguish the system are not made explicit. Individual organisms, at least those of moderate size, are naturally viewed as entities distinguishable from their environment. While it is somewhat less natural to view species as causal entities, they do appear naturally as classificatory units. Even without theoretical or empirical support for treating them as units, we readily group together various organisms on grounds of similarity into units closely resembling those associated with biological species. However, treating species as *causal units* represents a significant conceptual advance, as it requires, at least in part, construing species not as classes but as units capable of affecting other units and being affected by them (cf. Ghiselin 1974; Hull 1976, 1978). Thus, segmenting systems in nature for purposes of developing mechanistic explanations often relies on theoretical considerations. In the last case discussed above—that of biological respiration—the identification of cells

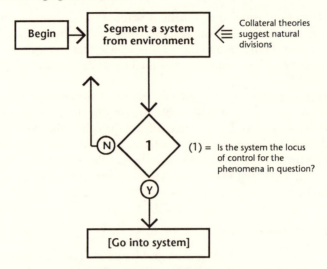

Figure 3.1. The Initial Choice. Defining a locus of control depends on segmenting systems of control from the environment and identifying systemic function. Identifying the locus of control and determining what qualifies as a system are correlative enterprises.

as structural and functional units was itself the product of scientific inquiry. Because the cell theory was already developed by the time some researchers on respiration sought to locate the phenomenon in the cell, it provided the necessary collateral theory.

Having segmented system from environment, one must then locate the locus of control within the system. An array of empirical and theoretical evidence is often brought to bear in settling on a locus of control for a phenomenon. As we have seen, some early investigators, under the influence of evidence suggesting that tissues could respire even without lungs or the circulatory system, concluded that tissues were the proper locus of control. The two most influential researchers opposing this claim, Ludwig and Bernard, tended to discount this evidence, but for quite different reasons. Ludwig produced evidence he thought showed that the circulatory system was really in control of respiration. Bernard thought the tissues and cells actually determined the rate of the reaction, but thought this was incompatible with them also being the locus of the reaction itself. To establish the cells as the locus of control Pflüger countered Ludwig's evidence, showing that cells are the controlling factor in the rate of reaction. He also proposed that cells could simultaneously be the site of the reaction and control its rate. He thereby justified the assumption that cells were a semiautonomous system in which one could situate the respiration process. Control was successfully localized.

The array of considerations invoked in discussing the locus of control for respiration were varied. Some were essentially theoretical or, perhaps, qualitative. Thus, the appeals to thermal differences and mechanisms of transport in debates over the site of respiration were largely qualitative. Bernard's exclusion of the tissues as the site of respiration rested primarily on theoretical considerations. In other cases the considerations are more narrowly empirical. Some of these are clear exemplars of localizationist experimental strategies, which we will see deployed in many different ways in subsequent chapters. Thus, Pflüger showed that respiration could occur even in the absence of blood to control it; variations in oxygen level did not effect respiration. Moreover, respiratory processes similar to those of animals occurred in insects, where blood could not serve a regulatory function. These empirical appeals are variations on the analytical techniques sketched in Chapter 2. Excitatory studies rely on enhanced activity with increasing stimulation. Varying oxygen levels should affect respiratory rate if it controls it. In Pflüger's case it does not. Inhibitory studies rely on diminished activity in the absence of a structure. Just as the inhibition of a function with the ablation of a structure implicates that structure, maintaining a function in the absence of a structure suggests it is relatively unimportant in the function; for example, if salt water will do, then hemoglobin is not necessary.

In the other two cases considered in this chapter there is still active disagreement. Those advocating internal control of behavior and evolution have tried to show that organisms or species do constitute semiautonomous units in important respects. Such researchers do not maintain that these systems are closed; rather, they acknowledge that in a variety of ways these systems are responsive to environmental factors, but they nonetheless defend an internal locus of control. For example, in arguing for internal control of language and conceptual systems, Chomsky does not deny that the environment influences the development of syntax and concepts, but he maintains that this effect is highly constrained by the internal operation of the system. The impact of the environment is, in his view, one of "prompting" rather than "controlling" behavior. Early advocates of orthogenesis maintained an analogous view, while more recent proponents adopt a much weaker opinion, holding only that the species' internal structure is *one* component with an independent contribution to evolutionary patterns. Function can follow form. Epigenesists, on the other hand, reject internal control, largely by trying to show how development or evolution is primarily ruled by external factors. They emphasize the pliability of the system in responding to these factors, thereby contending that any explanation of the system *must* focus on these factors external to it. The result of a negative verdict on the locus of control is the rejection of the proposed segmentation of nature. What is then needed is a new definition of the system that might be construed as the locus. Thus, the task for the epigenesist, as well as for the behaviorist, is the identification of a more comprehensive system. Only when that is done are they in position to proceed to the next step of inquiry by entering into the system itself to determine *how* it functions.

CHAPTER 4

Direct Localization

1. INTRODUCTION: RELOCATING CONTROL

Having "isolated" a system and identified it as a locus of control, the next step is to ask how the system does what it does. The goal is one of identifying and elaborating the mechanisms underlying behavior. A variety of approaches are possible. One that is often employed—especially in the earliest stages of a research program—is the identification of some component of the system that is itself responsible for producing the behavior, still leaving aside the question of how that component produces the effect. This is what we have called direct or simple localization. Responsibility for an effect is localized directly in a single, constituent component of the system. A system with complex capacities must have a complex structure. Direct localization acknowledges this and purports to explain complex capacities as a multiplicity of capacities. That is, direct localization proposes an analysis of the system into a set of components, each responsible for a specialized capacity. With direct localization, interaction among components is assumed to be either insignificant or nonexistent.

The empirical evidence for localization lies in the observed behavior of the system. In some cases, this is no more than correlational evidence; in others, we have experimental intervention in the form of excitatory or inhibitory studies (cf. Chapter 2). However, in each of the cases we will discuss in sections 2 and 3, independent lower-level constraints play a minimal role in the resulting models. In this sense direct localization is localization subject to minimal empirical and theoretical constraints—or, as we will sometimes say, it is *minimally constrained localization*.

Direct localization is prima facie reasonable, if only as a first approximation. Suppose we are confronted with a device that has an unknown structure and a complex output. If we propose to explain how some aspect of its behavior is controlled, one natural question to ask is: What part of the system controls it? We assume that the device is decomposable and that some specific component is responsible for its various behaviors. The practice is common with human institutions, as when we ask what person in a corporation is responsible for a given decision. Likewise, we may ask what part of a computer allows us to store information.

Identifying a component in the system responsible for a particular effect differs from identifying the locus of control *only in the level of analysis*.

Determining the locus of control requires segmenting a system from its environment and ascertaining whether control is internal or external. Direct localization requires segmenting a system into components and isolating some component or components within the system as controlling an effect; it transfers the locus of control for some dimension of system behavior to a lower level of organization and localizes control in a component subsystem. The proximate environment for the subsystem is the rest of the system.

Just as with the identification of a locus of control, direct localization involves a number of important and contentious assumptions. Jointly they are tantamount to assuming the decomposability of the system. Direct localization assumes that there are a number of components in the system, that these components function independently, and that any complexity in the behavior of the system is the effect of isolable subsystems. From the perspective of the behavior of the system, this entails that the various dimensions in its behavior are also relatively independent. Granting these assumptions, direct localization identifies component systems and proposes that one or several parts constitute a lower-level locus of control for each dimension of system behavior. Empirically this requires correlating changes in the behavior of component subsystems with changes in gross behavior, and then showing that changes in the former explain the latter.

In some cases the needed correlations are readily gained without elaborate intervention. This may mean searching out straightforward correlations between traits, as when we find that there is a correlation between smoking and lung cancer, or between performance on IQ tests and scholastic performance.[1] In other cases obtaining the correlations may require moderate intervention, as when we find that there is a correlation between REM sleep and dreaming. Similarly, we may investigate neuroanatomy by magnetic resonance imaging or regional cerebral blood flow in order to ascertain functional properties of the cortex. Finding the correlations may require substantial statistical analysis, as when we use statistical manipulations to separate signal from noise in isolating the activity of single neurons, or when we determine the characteristic signature of regions of the brain in EEG readings. In yet other cases, obtaining the needed correlations requires sophisticated experimental intervention. Finding the correlations may rely on naturally or artificially induced abnormalities, as when we find that there is a correlation between schizophrenia and an excess of dopamine in the central nervous system, or we realize that hemophilia affects only males. In these cases there are a variety of experimental techniques to identify component subsystems that control or modulate system behavior. In one way or another we can alter the behavior of a subsystem and then look for associated changes in the behavior of the system as a whole. As we pointed out in Chapter 2, one common experimental procedure is to disrupt the performance of the subsystem and then show that there is a corresponding inhibition of the effect one wants to explain. Another alternative is to intensify the activity of the subsystem and then show a corresponding increase of the effect in question. The use of such inhibitory and excitatory techniques is as controversial and difficult as it is common.

Direct localization, if correct, does explain why the system *as a whole* behaves as it does. We shift to a lower level of organization, identifying component parts and organization; then we isolate the component, if there is one, that is responsible for what the system does. If we want to know why an individual is schizophrenic, the presence of elevated dopamine levels is both relevant and important. Direct localization does not, however, provide an ultimate explanation, as it does little more than locate an underlying system within a complex system. Even if direct localization is successful, it tells us only *what* produces the effect, and not *how* it is produced. We may still want to know how the component system produces the effect it does, but that question is deferred rather than answered in seeking a direct localization. More positively, direct localization relocates the problem as one to be answered by research geared to a lower level of organization.

Even though direct localization does not produce a full mechanistic explanation, use of the heuristic is often an important preliminary step in the search for mechanisms. Without isolating a component that exerts control, if there is one, we cannot begin to provide a detailed mechanism. If control turns out to be more complex, involving several components, finding a relevant component may facilitate the search for others. Direct localization is a first pass at parsing the behavior and control of a system into causally significant segments. It is also a common strategy. Whether the resulting model of the system and its organization is right or wrong, this is a strategy that a realistic model of discovery would be ill-advised to neglect. In the sections that follow we will sketch several models incorporating direct localization. In section 2 we turn to phrenology; in section 3 we examine two competing accounts of respiration—those of Heinrich Wieland and Otto Warburg.

2. Phrenology and Cerebral Localization

In the nineteenth century Sir William Hamilton described phrenology as "idiotcy grafted upon empiricism" (cited in Cooter 1984, p. 48). Adam Sedgwick was at least as disparaging, describing it as a "sinkhole of human folly and prating coxcombry" (cited in Young 1970, p. 10). More recent accounts often follow the lead of Hamilton and Sedgwick, treating phrenology as a pseudoscientific enterprise, comparable to the use of water cures to treat recurring ills, reflecting nineteenth-century social fads and fashion, but hardly deserving treatment as a serious scientific view. This view has no credibility in light of historical treatments of phrenology by, among others, John D. Davies (1955), Robert M. Young (1970), Barry Barnes (1974), David de Guistino (1975), Stephen Shapin (1975, 1979), and Roger Cooter (1980, 1984). Though in the later phases of the movement phrenology was marked by a tendency toward popularization, it is a caricature to treat it as merely a piece of "Continental quackery," as it was often described. Moreover, it is important to recognize that the dichotomy between science and pseudoscience is often little more than a rhetorical flourish designed to enforce a particular point of view (cf. Mendelsohn 1977; for a contrary view, see Thagard 1988, ch. 10)—an ideologically conservative view in particular (cf. Cooter 1980). The actual opposition to phrenology was largely grounded in the antimaterialist and antispeculative movements of the early nineteenth century. In France, under the leadership of Georges Cuvier and the explicit assaults by Flourens and Magendie, phrenology hardly had a hospitable climate. In Britain it fared little better. Indeed, the phrenological societies that flourished in Britain during the early decades of the nineteenth century can be seen as a response to the exclusion of phrenological research from the Royal Societies and the British Association for the Advancement of Science, which were under the conservative leadership of William Whewell and Adam Sedgwick.[2] Despite the parody of phrenology promoted by its opponents, Franz Joseph Gall (1758–1828) and Johan Gaspar Spurzheim's (1776–1832) emphasis was clearly empirical, even though it was based on comparative anatomy rather than experimental ablations or the study of neurological deficits.[3]

In understanding phrenology and its pretensions it is useful to understand its historical context, and particularly its debt to one strain of Cartesianism. Cartesian physiology established that the nervous system was to be understood in mechanistic terms: its functioning is mechanical and automatic, requiring the intervention of no conscious agent. Cartesianism also bequeathed to us the doctrine that the brain is "the organ of the mind": all action and perception is mediated by the brain, and it is only through this organ that the mind influences the body.[4] According to this view the soul no longer animated the entire body, but rather exerted its influence through the brain. The overall effect of the Cartesian orthodoxy was no doubt salutary for physiology, as it freed physiological investigation from the vitalistic assumptions of its predecessors; however, the brain, though mechanical in operation, was regarded as a single functional unit, largely or wholly undifferentiated as to function. The mind was taken to animate the brain as it once had animated the body. At the same time, Cartesian metaphysics exempted the actions of the soul from the domain of empirical science; the Cartesian mind was nonmechanical in its operation. The investigation of *cognitive* activity—of the "higher functions"—became a matter mainly for introspection and speculation.

The assault on the citadel of Cartesianism did not begin in earnest until the nineteenth century. Gall led the break with the tradition by maintaining that the brain is *not* a homogeneous unit, but consists of a variety of organs, or centers, each subserving specific intellectual and moral (that is, practical) functions. These organs were localized in the cerebral hemispheres and thus were distinct from those organs subserving both the vital functions and the affections, which he localized in "lower" portions of the brain. The various intellectual and moral functions were, in turn, diverse from one another and were likewise localizable in discrete portions of the cerebral hemispheres. Gall's more infamous commitment to craniology was, of course, coupled with this view of the brain as a network of relatively independent organs. He held that the skull was a rather malleable structure which responded to the growth of the underlying nervous tissue and, accordingly, could be used as an index of the size of the underlying organs. Since he believed that the size of the underlying organs determined the extent of the respective intellectual abilities, the shape of the skull could be used as an empirical measure of the intellectual abilities (see Figure 4.1). Craniology was the source of phrenology's popularity among Gall's contemporaries, as it is the source of its derision among ours. Yet Gall's enduring contribution was his organology. It was also the focus of the main controversies surrounding phrenology among his contemporaries, as we will see in Chapter 5.

Phrenology was carried to England by Gall's collaborator, and the person who also coined the term *phrenology*, Spurzheim. Spurzheim made minor modifications to the organology (much to Gall's dismay), and it eventually gained serious attention in both the United States and Great Britain.[5] At the height of its popularity in the 1830s, the phrenological movement encompassed between forty and fifty independent phrenological societies in the United States alone. Under the leadership of George Combe in Britain and Orson Squire Fowler in the United States, it left its mark not only on medicine but on education and politics as well.[6]

In any of its incarnations and variations there is a tripartite commitment at the heart of phrenology: the cerebral hemispheres are held to be a collection of relatively independent organs which subserve the primary intellectual functions; the size of these respective organs is correlated with, and the basis for, differences in abilities; and, finally, the size of these organs, and thus the differences in abilities they reflect, can be measured by examining the shape of the skull. The capacities of individuals are explained as the products of organs localized in the cerebral lobes,

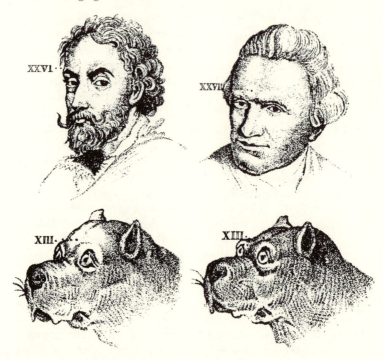

Figure 4.1. Phrenological Faculties and Cranial Localization. These cranio-logical representations are reproduced from an American reprinting of Spurzheim's *Phrenology, or the Doctrine of the Mental Phenomena* (1833). It is important that phrenological analyses incorporate both interspecific and intraspecific comparisons. What are called *affections* tend to be poste-rior, including, for example, propensities toward combativeness or secre-tiveness, as well as sentiments such as self-esteem and benevolence. The intellectual powers are anterior, including such things as senses of time or language.

though the activities of these organs were not themselves subjected to examination or explanation. Together these assumptions embodied a re-search program intent on transforming psychology into a more "biologi-cal" study and one that was less speculative than its competitors. This was true of the craniology no less than the organology. As Young explains, Gall "was wrong, but his [cranioscopic] hypothesis was extremely plausible at the beginning of the last century, and it played a very important part in the transition from speculations about unverifiable physiological homoge-neity to the experimental study of the brain" (1970, pp. 14–15). Let us look briefly at each of the three phrenological commitments and their contributions to the program of phrenology.

First, it was emphasized that the exercise of psychological faculties depended on their physiological realization: "The faculties and propensities of man have their seat in the brain" (Gall 1810–1819, 1:10). Spurzheim later described this as the "First Principle of Phrenology": "The brain is the organ of the mind" (1832, p. 6). In support of the principle, Spurzheim marshaled a variety of what were by then commonplace observations: the brain is necessary for feeling and thought; defective manifestations of thought accompany defective organization; and healthy development of the brain suffices for "energetic affective and intellectual powers" (ibid.). Together these were taken to imply the dependence of mental functioning on brain development. Phrenologists differed among themselves over the kind of organization to be found in the brain and the relation between mental faculties. Gall tended, for example, to deny any interaction whatsoever between faculties, while Spurzheim thought interaction was an important feature in explaining behavior.[7] None, however, challenged phrenology's "First Principle."

Second, and most centrally, phrenologists maintained that there was a correlation between the activities of the several faculties and the size of the appropriate physiological organs. This was Spurzheim's "Second Principle of Phrenology": "The mind manifests a plurality of faculties, each individually by means of a particular organic apparatus" (1832, p. 10). O. S. Fowler, the most prominent of the American phrenologists, ties the first two principles together:

> Phrenology points out those connexions and relations which exist between *the conditions and developments of the brain, and the manifestations of the mind,* discovering each from an observation of the other. Its one distinctive characteristic feature is, that each class of mental functions is manifested by means of a given portion of the brain, called an organ, the size of which is the measure of the power of the function. (1848, p. 5)

There was considerable variation concerning exactly how many and what these faculties were (for one taxonomy of the faculties, derived from Spurzheim, see Table 4.1), but it was agreed that the *faculties* were the species of affection or intellection and that the *organs* were the means by which the faculties of the mind were made manifest. A specialized apparatus for each function seemed to Gall and his followers to be a natural corollary of an analogy between the brain and organic structure. As George Combe, the leader of the Edinburgh phrenologists, said,

> Any theory, founded upon the notion of a single organ, is uniformly at variance with all that is ascertained to be fact in the philosophy of mind; . . . on the other

1. Affective Faculties
  1.1 "Propensities," the desires due to instinct
      1.1.1   Vivativeness, or the instinct for survival
      1.1.2   Alimentiveness, or the appetite for food
      1.1.3   Destructiveness
      1.1.4   Amativeness, or physical love
      1.1.5   Philoprogenitiveness, or caring for offspring
      1.1.6   Adhesiveness, or bonding (friendship)
      1.1.7   Inhabitiveness, or the tendency to be attracted to specific types of localities
      1.1.8   Combativeness, or the tendency to fight
      1.1.9   Secretiveness
      1.1.10 Acquisitiveness
      1.1.11 Constructiveness
  1.2 "Sentiments," which are not confined to inclination
      1.2.1   Cautiousness
      1.2.2   Love of Approbation
      1.2.3   Self-Esteem
      1.2.4   Benevolence
      1.2.5   Reverence
      1.2.6   Firmness
      1.2.7   Conscientiousness
      1.2.8   Hope
      1.2.9   Marvelousness
      1.2.10 Ideality, or enthusiasm
      1.2.11 Mirthfulness
      1.2.12 Imitation
2. Intellectual Faculties
  2.1 External Senses
      2.1.1   Feeling, as in pain, pleasure, temperature, etc.
      2.1.2   Taste
      2.1.3   Smell
      2.1.4   Hearing
      2.1.5   Sight
  2.2 Perceptive Faculties
      2.2.1   Individuality, producing the idea of being or existence
      2.2.2   Configuration, producing knowledge of patterns
      2.2.3   Size
      2.2.4   Weight
      2.2.5   Coloring
      2.2.6   Locality, producing knowledge of "the relative localities of external objects"
      2.2.7   Order
      2.2.8   Calculation
      2.2.9   Eventuality
      2.2.10 Time
      2.2.11 Tune
      2.2.12 Language, or the "power of knowing artificial signs"
  2.3 "Reflective" faculties, which operate on other sensations and notions; constitutive of Reason.
      2.3.1   Comparison
      2.3.2   Causality

Table 4.1. The Varieties of Mental Powers According to Spurzheim.

hand, the principle of a plurality of organs, while it satisfactorily explains *most* of these facts, is consistent with *all* of them. (1835, p. 19)

Discrete functions and abilities require distinct organs. This isomorphism, however, was explicitly restricted by phrenologists to "fundamental" or "primitive" faculties—those that maintain substantial independence from one another. In Gall's view the flaw in speculative faculty psychology was the failure to focus on these faculties. This was, according to Gall, also the source of the failure to recognize cerebral localization. As Spurzheim said, "The essence . . . of every faculty is always perceptible, . . . the essential nature of each primary power is invariable, and no organ can produce two species of tendencies" (1832, p. 21). The empirical implication is that there are a number of faculties between which there is relatively little correlation. For example, the sense of taste may be highly developed though the sense of hearing is not; similarly, the tendency toward destructiveness may be lacking while that toward combativeness is not.

Third and finally, the skull reflected the form of the brain. Spurzheim described it this way: "From birth and through mature years, up to the period when the faculties fall into decay, the size and form of the brain and its parts may be determined by the size and form of the external head" (ibid., p. 15). In Gall's view, any "difference in the form of heads is occasioned by the difference in the form of the brains" (1810–1819, 1:55). This final phrenological commitment provided the empirical criterion for localization. Phrenologists sought correlations between cranial structure and behavioral propensities, thinking that the only reasonable explanation for the correlations would be based on increased development of the appropriate brain structures. This study included comparisons of individuals: those having some pronounced or unusual capacity were examined for peculiar or extraordinary prominences, and those lacking capacities were inspected for the lack of corresponding cranial structures.[8] The craniological assumption was important to phrenology as an operational criterion. Because cranial structure was supposed to reflect brain structure, any correlations between cranial structure and special abilities was taken to be significant.

Organology was central to phrenological thought; it constituted an assumption of discrete faculties and their localization in correspondingly discrete regions of the cortex. As Herbert Spencer wrote, "A FUNCTION to each organ, and each organ to its own function, is the law of all organization" (1851, p. 274). In the simplest of cases, exemplified in Gall's writings, it was assumed that the organization was aggregative. Gall assumed that there was no significant interaction between faculties and, accordingly, that complex abilities were simply aggregates of simple abilities. Spurzheim's most significant departure from his teacher's views was precisely in the form of organization he allowed. On the one hand Spurzheim emphasized that many effects are due to "the mutual influence of the primitive faculties" (1832, p. 13). In his hands the organization was no longer aggregative, and though the primitive faculties maintained discrete localization in the brain, they were no longer functionally distinct.[9] The relative simplicity of either form of organization guaranteed that, as we shift to a lower level of organization in looking at the brain, we could transfer control for any basic cognitive capacity to a discrete unit at the lower level. Gall, on the other hand, supported the view that we have wholly independent organs with independent capacities. This was direct localization in its simplest and starkest form.

3. Competing Models of Cellular Respiration

We now turn to two theories involving attempts to localize cellular respiration, or biological oxidation, in different biochemical components of the cell. By focusing on an ongoing dispute between two theories within a research tradition that presupposed direct localization, we can see how such controversies depend upon common assumptions. This dispute differs from the previous one in crucial ways. Unlike the phrenologists, researchers in biochemistry did not rely simply on an observed correlation; they also developed experimental techniques to evaluate the correctness of their claims to localization.

Pflüger's research established that respiration occurred within the cell. This was not in itself an attempt to explain how respiration was accomplished. An explanation was needed, though, since the substances involved—carbon, hydrogen, and oxygen—do not readily react at the temperatures found in the cell. Intracellular respiration therefore appeared anomalous. What was required was an account of how the cell was able to facilitate respiration; that is, the problem was how to explain the reactions involved in oxidation given the intracellular temperatures. Researchers assumed that there was a simple mechanism, a catalyst, that facilitated oxidation. The problem then, became one of identifying and localizing this catalyst. Investigations into processes such as fermentation had led to the introduction of the term *enzyme* for a catalyst found within a cell.[10] Because the enzymes were identified in terms of their specific effects, the central differences between competing models were to be found in what activities or functions the accounts proposed for these enzymes.

Attempts to explain biological oxidation in chemical terms began in the nineteenth century using two very different approaches. One, by Schoenbein and Traube, focused on an alteration of molecular oxygen to make it more likely to react with the substrate. Schoenbein's proposals stemmed from his discovery of ozone. In 1845 he showed that ozone could oxidize guaiacum, and in 1848 he showed that potato roots had the same effect. From these results he constructed a general theory in which the formation of ozone, which he took to be an especially active form of oxygen, was an intermediate step in biological oxidation. Traube also developed an account that focused on oxygen, but instead of proposing a specialized form of oxygen he proposed a reaction in which ordinary oxygen was transported to a substrate. In Traube's models this typically involved the formation of hydrogen peroxide from water and molecular oxygen, which in turn would oxidize the substrate. In contrast to both of these models, Hoppe-Seyler developed a model according to which the crucial step was the formation of molecular hydrogen, which in turn possessed unusual capacities to react with oxygen (for more detail, see Kastle 1910).

These nineteenth-century proposals constitute direct localizations in that they propose a single mechanism for explaining biological oxidation. Each of the models postulated single enzymes with specific functions, based on an understanding of cellular function. However, because of limitations on experimental techniques, it was difficult to elaborate on these models. Each, moreover, faced serious empirical and theoretical problems. We turn now, though, to two analogous models advanced and developed in active research programs in the early twentieth century. These differ from each other fundamentally on their understanding of the action of the enzymes in catalyzing the oxidation reactions of cellular respiration. Wieland, later supported by Thunberg, argued for a process in which enzymes removed hydrogen from substrates. The key to the proposal—one reminiscent of Hoppe-Seyler—is the enzyme that removes the hydrogen from the substrate, thereby allowing it to react with oxygen. The alternative model, by Warburg, incorporates *oxygen activation*, according to which oxygen, once activated on the surface of his proposed agent, the *Atmungsferment*, would react directly with the substrate.

These alternative models share a simple structure. Each posits a single enzyme with a specific effect. Since each enzyme could, in principle, account for the net effect of respiration, experimental techniques were introduced to determine the detailed conditions under which respiration occurred. There was heated controversy surrounding this issue until the mid-1920s. We now know that both Wieland and Warburg identified an important functional component that figures in cellular oxidation. They both, however, attributed responsibility for the total oxidation to a *single* component, when in reality *each* component figures as a constituent function in the overall process; and neither researcher was willing to consider a more complex organization. It was not until the 1920s that investigators began to appreciate how the two accounts could be viewed as identifying components of a much more complex process; and it was only in the 1930s that the basic architecture of this more complex process was identified. Our focus in the remainder of this section will be on the period prior to reconciliation and on the arguments that were used to support the competing models of Wieland and Warburg.

Wieland's Localization of Respiration in Dehydrogenases

Wieland's initial research focus was not on biological oxidation, but on oxidations performed by inorganic catalysts like palladium black, a reaction first studied in the early nineteenth century by Humphry Davy. Wieland studied this catalysis, showing that when no oxygen was present, palladium could remove hydrogen from many compounds. This reaction, however, could only proceed for a short time—until the palladium was saturated with hydrogen. This suggested that oxygen was required *only* to receive the hydrogen released from the substrate by the palladium. It was important, but only secondarily so. To demonstrate this Wieland employed methylene blue (a readily reduced dye), showing that when it was available to receive the hydrogen from the palladium the reaction could continue even in the absence of oxygen. Wieland referred to both the oxygen and the methylene blue as simply "hydrogen acceptors," indicating that they played a basically passive role in the reaction. He thus contended that oxidation consisted, in such cases, not in the addition of oxygen to the substrate, but in the removal of hydrogen:

> By this method of considering the matter, the catalytic action of the platinum or palladium in these processes does not occur by the metal activating a molecule of oxygen (with intermediate formation of a peroxide), but much more probably by the metal activating hydrogen, as is believed to occur in the purification of detonating gas. (1913, p. 3327)

Thus, Wieland proposed the following general scheme for oxidation of a substrate (R) by palladium (Pd), with methylene blue (Mb) serving as the hydrogen acceptor:

$$RH_2 + Pd \Rightarrow R + PdH_2$$
$$PdH_2 + Mb \Rightarrow Pd + MbH_2$$

A number of other oxidation reactions, however, seemed more plausibly construed as involving the *addition* of an oxygen atom to the substrate. For example, the overall reaction in the case of oxidizing an aldehyde can be represented as

$$RCHO \Rightarrow RCOOH.$$

Aldehydes do not possess two hydrogen atoms to be removed. Wieland argued, however, that these reactions still followed the same scheme and required an intermediate process of hydration. He thus proposed the following for the oxidation of an aldehyde:

$$RCHO + H_2O \Rightarrow RCH(OH)_2 \Rightarrow RCOOH + H_2$$

Wieland similarly suggested that the oxidation of carbon monoxide involved an initial hydration to form formic acid ($CHOOH$), followed by dehydrogenation to leave carbon dioxide. He maintained that *all* oxidations consisted simply in the removal of hydrogen and did not necessarily involve oxygen.

The contention that oxidations were processes of dehydrogenation prepared Wieland to draw a connection between oxidation and reduction. These processes had previously been conceived as independent, with oxidation involving the uptake of oxygen. In Wieland's scheme, though, reduction and oxidation became "two expressions of one process of dehydrogenation" (ibid., p. 3340). The hydrogen acceptor, whether oxygen or methylene blue, was reduced by the hydrogen removed from the substrate being oxidized:

> If we consider oxidation processes as dehydrogenations, as the foregoing results have indicated exactly, at least for some important cases, then we have a reduction process at the same time, since the hydrogen activated by the ferment must be taken up by some sort of acceptor. (Ibid., p. 3339)

It would follow that no substance can be oxidized without another being reduced. According to Wieland's model, the oxygen that is added to the substrate in oxidation, or contained in the carbon dioxide or water released, does not come from molecular oxygen, but rather from either the original substrate or water, and is of only secondary importance.

Wieland then took up biological oxidations to show that they could be accounted for in a similar way. This extension of models of reduction to oxidation makes Wieland's work a particularly interesting case of direct localization. Reduction was thought to be a chemical process. By treating oxidation as an analogue of reduction, Wieland sought a chemical explanation for what was then thought of as a biological process. Wieland's strategy was to posit an agent in the cell that played the same role as palladium black in the purely chemical oxidations he had studied. He intended to show that biological substrates, such as grape juice, could also be dehydrogenated—that is, oxidized—with palladium black, even in the absence of oxygen, provided another substance, such as methylene blue, subsequently removed the hydrogen from the palladium. Moreover, he showed that the reaction began, even without methylene blue, "with a rich formation of carbon dioxide." The reaction predictably slowed as the metal became saturated, and resumed when methylene blue was supplied.

Recognizing that "this is still only an imitation of the biological oxidation process when it is obtained with a substance foreign to the cell as the catalyst," Wieland demonstrated that organic ferments could also function with hydrogen acceptors other than oxygen—again, most commonly, methylene blue. He thus showed that bacteria could oxidize ethanol and acetaldehyde to acetic acid if oxygen was lacking but methylene blue was provided. The fact that this reaction could occur without oxygen convinced Wieland that biological oxidations involved the same process—the removal of hydrogen from the substrate—as did chemical oxidations with palladium. Oxygen was inessential. Wieland restated his basic argument in *On the Mechanism of Oxidation*:

> As it can be proved that one and the same enzyme system uses in the one case molecular oxygen, and in the next a quinone derivative to fix the hydrogen which has been removed, it follows that the reaction mechanism must be similar in both cases. Common to both is hydrogenation of an unsaturated molecule, namely oxygen or quinone. This uniformity in mode of action cannot be attributed to an activation of the oxygen, but is satisfactorily explained by the assumption that the hydrogen in the substrate is activated. (1932, p. 30)

Wieland was not able to identify directly the catalytic agent he claimed was involved in oxidation within living cells. These agents could only be identified functionally in terms of the reactions they catalyzed. Since he could not isolate and study the enzymes, he had to proceed indirectly and examine the overall reactions occurring in the cells under a variety of circumstances. The circumstances that proved critical were those that showed oxygen was not necessarily involved in respiration and that water was required for some oxidations. In addition to studies of this form, Wieland tried to show the similarities between his model system involving palladium black and the biological cases. Thus, he attempted to determine the kinetics of milk dehydrogenases and study the role of hydrogen peroxide in these reactions in order to compare them to his model systems (ibid., ch. 4). The parallels between the behavior of biological systems and the predictions he made on the basis of his dehydrogenation model were, he thought, convincing evidence that hydrogen rather than oxygen was activated in these reactions.

Except for one critical response by Bach (1913), Wieland's proposal did not initially attract much attention, negative or positive. One factor that helps to explain this is Wieland's role as an outlier to the emerging biochemical community. He was not principally concerned with reactions in living systems, but more generally with organic chemistry.[11] Accordingly, Wieland's approach to oxidation was more typical of an organic chemist; his focus was principally on the structural changes that occur in oxidation.

Moreover, he was trying to oxidize organic compounds using palladium as a catalyst. When he attempted to extend his theory to biological oxidations, he concentrated on tracing the kinetics of the biological reactions and showing that they corresponded to the kinetics of reactions achieved in his model system. As a result he did not construe the problem in the same way as had more biologically oriented researchers, who focused on the enzymes potentially involved and tried to identify them either through purification or inhibition.[12]

There was already recorded evidence in physiological chemistry that could be interpreted to support the idea that oxidation occurs through dehydrogenation. Wieland, however, did *not* establish the auxiliary connections; they were developed a few years later by Torsten Thunberg, who saw the relevance of Wieland's scheme to the biological research he was conducting. Battelli and Stern (1911) had worked with the oxidation of succinic acid. In accordance with the view that oxidation involves a direct reaction of a substrate with oxygen, they took the product of this oxidation to be malic acid. Einbeck (1914) showed, though, that the first step in oxidation was the formation of fumaric acid. The reactions can be represented as follows:

$$HOOC-CH_2-CH_2-COOH \Rightarrow HOOC-CH=CH-COOH$$
$$\text{[succinic acid]} \qquad\qquad \text{[fumaric acid]}$$

$$\Rightarrow HOOC-CH_2-CH(OH)-COOH$$
$$\text{[malic acid]}$$

The intermediate step is curious from the perspective of traditional theories of oxidation, but it makes perfect sense from the perspective of Wieland's dehydrogenation theory: fumaric acid is dehydrogenated succinic acid; malic acid is formed by hydration of fumaric acid.

To defend his account, Thunberg required not just an interpretation of data in accord with Wieland's theory, but experimental evidence showing that the biological reactions did indeed go through the kind of hydration and dehydrogenation Wieland's theory required. Biochemists were already in the process of developing experimental tools for analyzing such stepwise reactions. Knoop (1904) and Dakin (1912) had proposed a scheme for fatty acid metabolism that involved a sequence of oxidations of the beta carbon.[13] Demonstrating this scheme depended on isolating the proposed intermediary substances. Since these substances tended to be metabolized as rapidly as they were formed, however, this proved difficult. Thus, techniques had to be developed to trap intermediaries or interrupt the process at certain stages to identify the intermediate products that were produced.

Thunberg's investigations stemmed from his observation that succinic acid—the final product formed in the oxidation of fatty acids with active oxidizing agents such as nitric acid—could be readily oxidized by living tissues. He constructed a device (later known as the Thunberg tube) which was evacuated of air. Into it he put muscle substance and methylene blue. When succinic acid was added to the tube, the methylene blue was rapidly decolorized. He concluded that the enzymes of the muscle tissue had activated the hydrogen in succinic acid (Thunberg 1916). He then needed to demonstrate the stepwise character of this oxidation. In his earlier experiments, Thunberg had found that addition of other substances could also increase oxygen uptake of tissues. He now tested numerous substances in the Thunberg tube and showed that malic acid, fumaric acid, citric acid, and lactic acid, among others, could all be dehydrogenated with methylene blue. Although Thunberg could not yet isolate the enzymes responsible for these effects, he *was* able to show that the capacity of the cell to carry out various of these reactions was differentially affected by heat. Knowing that different degrees of heat could destroy or inhibit catalytic capacities of enzymes, he argued that there were discrete enzymes responsible for the different reactions (1920; for a review, see Thunberg 1930).

Thunberg's ability to incapacitate selective enzymes responsible for different steps added critical support to Wieland's claim that oxidation was carried out by specific localized agents in the cell. In particular, Thunberg saw this as evidence against the view that the cell structure as a whole explained the ability of cells to carry out these reactions—a view he saw as a vestige of nineteenth-century protoplasm theory. Thunberg allowed that cell structure could effect the harmonious linking of reactions and might play a role in "the transformation of chemical energy into other forms of energy." He contended, though, that the ability to carry out oxidation with destroyed cells counted against the need for such structure for the basic reactions:

> The dilatory conditions to which muscle mass has been subjected speak against the supposed importance of the cell structure to the reactions in question. The mincing of the muscle cells and the treatment with distilled water are factors which must derange the polyphasic system of the muscle cells, i.e., their structure. (1930, p. 326)

He maintained that a highly localized agent, one that was not disturbed by destroying cell structure but which was destroyed by heat, was the true agent of such oxidations. Thunberg's argument paralleled Wieland's in the claim that oxygen was not needed for oxidation. Both cases relied on showing that the reaction continued even when the supposedly critical factor (cell structure or oxygen, as the case may be) was removed. The central factor must therefore be something else, or so they argued.

In positing the explanation of oxidation by hydrogen activation, the Wieland-Thunberg scheme is a straightforward case of direct localization. Thunberg's account is more complex than Wieland's in proposing a chain of such oxidations in the cell, each due to a separate enzyme. Every step in oxidation is attributed to a single, specific agent, which executes its function in the cell by removing two hydrogen molecules from the substrate and then uniting the hydrogen with available acceptors. Thus, a direct localization is offered for each oxidation in the cell. In Thunberg's scheme, complex reactions are not explained in terms of an integrated system of reactions, but are decomposed into a series of simple and self-contained steps, which are linearly ordered and mediated by single, specific agents.

Neither Wieland nor Thunberg offered any independent evidence regarding the nature of the enzymes mediating hydrogen activation. Both submitted models based on the behavior of cells or cell extracts; neither suggested chemical mechanisms to explain how the proposed enzymes catalyzed cellular respiration. As we have said, an explanation of this sort is incomplete even when it is correct. Without some account of how the enzymes are capable of performing the dehydrogenations attributed to them, the scheme is not yet fully mechanistic. In the 1920s Wieland did finally propose such an account:

> It may be considered that partial valencies are concerned in the union of adsorbing and adsorbed substances and that the utilization of these partial valencies causes a readjustment of the valency relationships in the molecule of the adsorbent. An unmistakable chemical reaction occurs during the adsorption of hydrogen on platinum or palladium, and it is accompanied by a very considerable evolution of heat. Here we see the effect of the saturation of the partial valencies of the hydrogen molecule. The adsorption compound might be expressed as follows, where the number of palladium molecules involved is not expressed.
>
> $H\ Pd_1\ \ldots\ Pd_n\ H$
>
> By the saturation of the partial valencies, the link between the two [terminal] hydrogen atoms is weakened, and in this we see the reason for the activation of the hydrogen. During hydrogenation, this complex adsorption compound reacts with the similarly adsorbed hydrogen acceptor whereby the active hydrogen gives up its valency linkage and performs the task prescribed for it by the thermodynamic conditions. The continuous removal of the product of hydrogenation from the catalyst by fresh hydrogen produces the endless repetition of the surface reaction and, accordingly, the catalytic effect. (1922, p. 515)

For hydrogen bound within molecules, Wieland proposed that whole molecules attached to palladium. These palladium-enriched compounds involved a reduced internal valency in the molecule, allowing the hydrogen to join directly with the palladium, and from there on with other acceptors. Wieland's suggestion was no more than a speculation concerning how catalysts *might* perform their functions. Developing an empirically motivated model was a task for the future.

Warburg's Localization of Respiration in Membrane Iron

An alternative, but equally localizationist, model for respiration and oxidation was offered by Otto Warburg in the period 1910–1915. Like many of his nineteenth-century predecessors, Warburg construed oxygen as the active agent in oxidation. In the earliest part of this period, he emphasized structure and developed a view that accorded with the then-emerging colloidal conception of respiration, which was markedly opposed to the enzyme-based models. He proposed that the oxygen was activated by a structure on the membrane surfaces within the cell. He called this surface agent the Atmungsferment, viewing it as a general catalyst for oxidation and not as more specifically involved in certain reactions.

Warburg's early departure from the main tradition of enzyme accounts, emphasizing organization rather than component analysis, is explained partly by his background and his principal objectives as a biochemist. Warburg's work focused on cancer. The continuous division of cancer cells led Warburg to examine the process of cell division. His closest affiliations were therefore with researchers in experimental cytology and embryology, such as Loeb and Verworn, rather than with the emerging orthodoxy in biochemistry, which focused primarily on the extracts that could be studied in isolation from living cells (Kohler 1973a). Some of Warburg's earliest experimental work was on sea urchin eggs, in which he investigated Loeb's claim that the nucleus was the locus of metabolic activity. He concluded, contrary to Loeb, that membrane surfaces rather than the cell nucleus were the locus of respiration. He established this by showing that alkaline solutions increased the rate of respiration. These solutions only affect the acidity of the membrane, not the acidity of the protoplasm. It followed, he thought, that membranes played the key role in respiration. In a contrary fashion, fatty acids and organic solvents decreased the rate of respiration. Warburg interpreted this as resulting from their effect on the membrane. By a combined use of excitatory and inhibitory methods, Warburg claimed to establish that membranes are the components responsible for respiration in the cell.

Warburg's model clearly incorporated a direct localization, but in appealing to a membrane surface rather than an enzyme, it was quite different from Wieland's enzyme model: it fixed a different locus of control. It also attracted a distinct group of adherents in the early 1920s. Kohler, for example, explained the attractiveness of Warburg's theory as due to growing disillusionment with the enzyme view:

> Despite a steady stream of new discoveries, biochemists grew less certain that the basic mechanisms of life were simple enzyme reactions. The persistent failure to isolate a pure enzyme raised doubts that enzymes were proteins or that proteins were definite chemical substances. The blossoming popularity of colloidal chemistry at the time provided an attractive alternative view. . . . By the 1920s enzymes were widely regarded as small organic molecules or metal ions adsorbed on an ill-defined mixture of colloidal 'carrier' proteins. This was the prevailing mood and beliefs of the period in which Warburg's *Atmungsferment* flourished, and Warburg's view was very much in harmony with it. (1973a, p. 174)

One reason for this growing dissatisfaction with enzyme models was that although cell extracts in isolation could perform many crucial biological reactions, they performed them more slowly than did whole living cells. The natural conclusion, as Warburg saw, was that the reduced respiration in extracts from liver cells was due to the destruction of the intact structure of living cells. It is *not* the presence of chemical constituents that explains biological processes, but cellular structure:

> The biologists who work with press-juices of organs have generally observed that the important cellular functions are no longer present in cell extracts. It was generally presumed that structure was a necessary condition for most biological reactions. Knowing now the importance of cell surfaces for expiation, such results are no longer surprising. (1910, p. 328)

As we will explain in Chapter 7, similar retardation of the rate of reaction was found when fermentation was carried out in cell-free extracts. Warburg, in fact, challenged the enzyme theory of fermentation, arguing that it neglected the role of cell structure. He argued that while the "ferments," or enzymes, did figure in such reactions, it was the adsorption of the ferment onto the cell membrane that played the crucial role in accelerating the reaction (1912).

While Warburg argued against localization of the causal factors responsible for physiological functions *in enzymes*, he nonetheless pursued a localizationist strategy—and, indeed, a *direct* localizationist strategy. This showed up in his experimental approach, which was directed to determining whether something was or was not *the agent* responsible for a process. Thus, he studied the effects of a number of narcotics such as ethyl urethane to determine how they interrupted cellular respiration. He showed that the more lipid-soluble the narcotics, the greater their inhibitory effect. He concluded that it was not the lipid phase but the solid particles that were responsible for the increased effectiveness of these narcotics.[14] Instead of treating the poisoning as due to an alteration of the overall physical state of the membrane, he now attributed it to the adsorption of the narcotic on the solid surfaces, where it blocked access to a physical-chemical unit.

Otto Meyerhof, then working in Warburg's laboratory, made an important discovery at this point: citric acid and tartaric acid halted respiration in sea urchins. These acids were known to find and remove iron molecules. That suggested to Warburg that iron embedded in the cell membrane might be the crucial factor in explaining respiration. The fact that adding iron salts increased sea urchin respiration further supported this suggestion. As a consequence, Warburg proposed "that the oxygen respiration in the egg is an iron catalysis; that the oxygen consumed in the respiratory process is taken up initially by dissolved or adsorbed ferrous iron" (1914a, pp. 253–54). He now proposed that Atmungsferment consisted of ferrous iron, adsorbed onto the inner surfaces of cell membranes. Once oxidized, the iron would convert to what Warburg termed the "ferric state" and would serve to oxidize molecules of substrate, which were also adsorbed onto the cell surfaces. Even as the specific character of his explanations changed, Warburg's strategy remained localizationist, despite his opposition to enzymatic views.

Experimentally, Warburg emphasized the interruption of oxidation and attributed the overall reaction to *whatever* was thereby affected. Having concluded, on the basis of experimental disruption, that ferrous iron was the agent of oxidation, Warburg sought a model system in which he could study the process. For this purpose he developed experimental paradigms in which activated charcoal or pyrolised blood (both containing iron as constituents) would catalyze oxidations. It was this model that led him to appreciate more fully the role iron played in his Atmungsferment, as it allowed him to demonstrate again the efficacy of citric and tartaric acids in decreasing the rate of oxidation. As a result of these excitatory and inhibitory studies, Warburg's localization of respiration in his iron-based Atmungsferment seemed assured.

Warburg's work was interrupted by military service during World War I (see Krebs 1981). After the war, Warburg returned to studying the iron-based oxidation enzyme, using charcoal containing iron constituents as a model. He also began to use hydrogen cyanide as a poison. This was significant because hydrogen cyanide behaved differently than narcotics. Unlike narcotics, dosages of hydrogen cyanide that reduced respiration to 50% of normal levels did not inhibit functions such as cell division. The more important fact in establishing the role of an iron-based enzyme in oxidation was that the inhibitory action of hydrogen cyanide could not be a result of being adsorbed on the catalytic surface. It was adsorbed much more weakly than narcotics. Warburg therefore concluded that the effect of hydrogen cyanide was the result of chemical action between the cyanide and the iron catalyst, which now appeared to function like an enzyme.

Warburg used the charcoal models in two ways. One was to defend the role of his Atmungsferment. By showing that the artificial model behaved like living tissues in oxidizing various amino acids and that it was affected by narcotics and hydrogen cyanide in the same manner as living cells, he gained substantial support for the theory that an iron-based enzyme was the active component in biological oxidation.[15] The charcoal models also gave him a means of studying the catalytic agent. By showing that the charcoal formed by burning sugar in the presence of silicates was inactive—as was similar charcoal formed by adding iron salts before heating—but that adding hemin before heating made the charcoal catalytically active, Warburg established that iron had to be combined with nitrogen to be active.

In 1924 Warburg presented his iron-based model of cellular respiration and proposed a mechanism for its action (cf. Warburg 1925). He proposed that through a reversible change in valency, the iron in the catalyst transported molecules of oxygen to the substrate:

> Molecular oxygen reacts with divalent iron, whereby there results a higher oxidation state of iron. The higher oxidation state reacts with the organic substance with the regeneration of divalent iron. (1924, p. 479)

In this paper Warburg defined an Atmungsferment as "the sum of all catalytically-active iron compounds in the cell" (p. 494). In succeeding years other research showed that a large number of enzymes were involved in cell oxidation; Warburg, however, retained his localizationist view, maintaining that *all* enzymes were components of his Atmungsferment and that laboratory evidence for their individuality was only an artifact of the experimental techniques those investigators employed. The basis of Warburg's argument was that the breakdown of the oxidation process as it occurred with extracted enzymes was different from that in poisoned cells. With extracts different substrates are oxidized, but "[i]f a fraction of the respiration is inhibited by carbon monoxide, the uninhibited respiration is in every respect the same as the inhibited fraction. . . . It follows that all iron atoms which transfer the oxygen in respiration, in one kind of cell, are linked identically" (1930, p. 357). Thus, Warburg was able to maintain the thesis that his Atmungsferment was *the* respiratory enzyme.

One of the problems Warburg faced in developing his model of oxygen activation was that he could not separate and isolate the Atmungsferment from the cell. This was partly due, he thought, to the low concentration of the enzyme in the cell, and partly to its instability. Thus, he saw the need for an alternative technique for studying the oxygen activating system:

> It seems therefore that the old difficulties of the enzyme problem render advance impossible—the instability of the enzyme and its nearly infinitely low concentration. Only a method which is independent of the quantity of the enzyme and which does not require its separation from the cell can overcome these difficulties. (1930, p. 349)

In the later part of the 1920s Warburg developed an additional research method for studying Atmungsferment that met these requirements: poisoning with carbon monoxide. Carbon monoxide reacted with Atmungsferment much like it reacted with hemoglobin, with the exception that it required a much higher ratio of carbon monoxide to oxygen in order to inhibit oxidation. Warburg (1927) observed that carbon monoxide inhibited the effect of iron. This provided additional confirmation of the importance of the Atmungsferment, but it led to even more significant developments. One of the important facts about carbon monoxide poisoning of hemoglobin is that it is light sensitive. Under the influence of light, carboxyhemoglobin tends to dissociate. Warburg showed that yeast poisoned with carbon monoxide showed a similar response to light. The fact that the Atmungsferment was poisoned by carbon monoxide, and that this poisoning was reversed by light, gave Warburg vital information about the identity of the Atmungsferment; it allowed him to differentiate it from other iron compounds by virtue of their different affinities to carbon monoxide and their different sensitivities to light.[16]

In addition to differentiating the Atmungsferment from other enzymes, Warburg used carbon monoxide to fingerprint the enzyme based on its absorption spectrum. He subjected poisoned tissue to light of different frequencies and tested to see how much the carbon monoxide poisoning was reversed. Warburg (1928, 1929) thus identified the absorption spectrum of the enzyme and showed that, while it resembled that of iron porphyrins such as hemoglobins, it was not identical to any known porphyrin.[17] While the new experiments Warburg developed in the 1920s were elaborate and provided powerful support for the existence of an iron-based component of the cell involved in oxidation, the logic of his argument remained much the same as before. The fact that he could increase or decrease rates of respiration by employing factors that affected this iron agent led him to treat it as *the factor* responsible for cellular respiration. Once again, there was no room for more than one agent, and no tolerance for complex organization.

The Conflict between Wieland and Warburg

The power of direct localization can be underscored by turning to how the principal researchers responded to each others' views. Wieland and Warburg each argued that they had identified the primary mechanism of oxidation and that the mechanism proposed by the other party was either not involved in biological oxidations or was an artifact of the experimental procedures employed. Each investigator had a considerable investment in his own explanation, and their differences were reinforced by disciplinary differences. Even when researchers came to accept the involvement of reactions involving substrate hydrogen and molecular oxygen, both principals refused to acknowledge the value of the contributions of the other and continued to emphasize their differences. It would, however, be overly simplistic to view this conflict as simply one of ego. We think it is also a product of their respective research programs, the reasoning, and the kind of evidence each researcher offered to support his thesis: Each assumed that the task was to find a single, discrete agent. Each relied on inhibitory methods. Each used an artificial model system to provide evidence for his localized mechanism. Artificial models, however, can only show that a process like that proposed *might* occur, but not that it is sufficient, or even that it *does* occur in isolation from any other processes. Wieland and Thunberg, as well as Warburg, did offer evidence to suggest that the modeled process actually occurs in living organisms, but, even if compelling, that does not establish that only that process is operative. Finally, both Wieland and Warburg relied on evidence that if the process they proposed was interrupted, oxidation ceased. Such inhibitory studies, however, can only show that the process is *necessary* to the overall reaction, not that it is *sufficient* to explain it.

To appreciate this point, let us briefly consider the responses Wieland and Warburg each presented to the other. When Wieland confronted Warburg's alternative theory, one focused on oxygen activation, he tied it with that of Engler and Bach, construing the role of ferric oxide in Warburg's scheme as analogous to that of hydrogen peroxide:

> The most generally accepted view is that iron compounds 'fix' molecular oxygen in peroxides of high oxygen potential and liberate it from these in an active form to the substances to be oxidized in the cell. Since iron is able only in the divalent condition to activate oxygen, catalysis by iron of biological processes is only possible where the substrate can reduce the iron in the postulated peroxide to the ferrous state. (1932, p. 24)

Wieland pressed two critical objections. One challenged the generality of Warburg's model; the other threatened its internal adequacy. Thus, Wieland pointed out that oxidation by divalent iron could not account for all known biological oxidations, and that the ways of extending the divalent iron theory to cover these cases (by assuming that the iron occurs in complexes) were implausible and ad hoc. He went on to say that the "hypothesis of the activation of oxygen by peroxide formation is *a priori* unsatisfactory, for it is difficult to reconcile the specificity of oxidizing enzymes with such a theory" (ibid., p. 25). Several substances oxidizable in vitro were not oxidizable by living cells—something Wieland thought could not be explained if all that was required for oxidation was an attack by activated oxygen.

Warburg's research made extensive use of hydrogen cyanide as a tool for inhibiting oxidation and as pointing to a special oxygen-activation mechanism. Wieland acknowledged that hydrogen cyanide affected oxidation with oxygen differently than with methylene blue, and this, he says,

> has led to the raising of serious objections to the theory of simple hydrogen activation. On this theory we could not foretell that the activated methylene blue would react practically unaffected by hydrogen cyanide, while the action appears impossible when oxygen is present. Thus it has been argued that the modes of action of the two types of acceptors are different. Quinone and the dyestuff are apparently susceptible to hydrogenation by labile hydrogen while molecular oxygen requires special activation which derives through some particular function of the enzyme. (Ibid., pp. 38–39)

Wieland claimed that this theory required a specialized "co-ferment" to activate the oxygen, but contended that "the assumption of such complicated relationships is unnecessary." He proposed instead to explain the phenomenon by postulating a greater affinity of hydrogen cyanide than oxygen for the active surface of the enzyme. He maintained that such a process could also explain the fact that in different organisms, different hydrogen acceptors were effective (ibid., p. 41).

Wieland took the differences with Warburg seriously, as was evidenced by the fact that the last two of the six chapters in his book (1932) were devoted to the catalytic effect of iron. He asked "whether our knowledge of the catalytic effect of iron in autoxidation may be reconciled with our conclusions drawn on grounds of the dehydrogenation theory, or if the two are entirely irreconcilable" (ibid., p. 86). Wieland contended that the process of oxidation achieved with iron catalysts alone is quite different from what occurs in the cell—a fact he attributed to the multiplicity of enzymes involved in biological oxidation. He then focused his attention on how iron might figure in biological oxidations. He considered two models. The first was Warburg's scheme according to which iron shifted from a divalent to a trivalent state, thereby activating oxygen, which reacted with the hydrogen from the substrate (B); for example,

$$2FeO + O_2 \Rightarrow Fe_2O_4$$
$$Fe_2O_4 + BH_2 \Rightarrow Fe_2O_3 + H_2O.$$

The alternative scheme invoked a dehydrogenation of a ferrous hydroxide, forming hydrogen peroxide; for example, with oxygen, we have

$$2Fe(OH)_2 + O_2 \Rightarrow 2FeO(OH) + HO-OH$$
$$HO-OH + As(OH)_2OH \Rightarrow OAs(OH)_2OH + H_2O.$$

Wieland rejected the first proposal—that is, Warburg's—on the grounds that adding ferric iron at low temperatures had little catalytic effect. Ferrous iron, on the other hand, exerted considerable effect until it was transformed into the ferric state, at which point the reaction slowed dramatically. In the process of transforming ferrous iron, though, the amount of oxygen used in the reaction exceeds the amount theoretically required. Wieland therefore defended the second model, involving ferrous iron catalyzing a dehydrogenation, as in better agreement with the oxygen-consumption data. He proposed that iron functioned in the ferrous hydride by activating the hydrogen. According to this model, the oxidation of the iron did not catalyze the reaction, but actually brought it to an end. Wieland thus tried to show that the iron Warburg claimed was involved in oxygen activation was in fact functioning in hydrogen activation.

Thus, even as late as 1930, Wieland remained dedicated to a single-factor model with direct localization. He viewed Warburg's mechanism as a competitor and as inconsistent with his own. Accordingly, he attacked the specifics of Warburg's model and reinterpreted Warburg's evidence to make it compatible with his own hydrogen-activation model.

Warburg, for his part, viewed Wieland's theory as highly speculative. He prized his own experimental ability and tended to view the work of other researchers as less exacting.[18] At first he simply denied that the dehydrogenation model was empirically viable. In the 1930s, however, his own experimental work led him to accept that oxidation did involve, in part at least, a mechanism for removing hydrogen from substrates. He identified coenzymes that figure in the transport of the removed hydrogen atoms. However, this did not lead to an acceptance of Wieland's earlier work, which he continued to reject as simply speculative chemistry:

> The oxidation theory of Wieland, no less than the opposing oxidation theories of Engler and Bach, were premature because when they were proposed nothing was known about the chemical constitution of the ferments participating in the respiration. They were theories regarding the mechanism of chemical reactions,

proposed without knowledge of the participants in the reaction. Such theories cannot be other than erroneous, and they must disappear to the extent that the chemical nature of the reaction partners—in this case the chemical constitution of the ferments—is elucidated. (Warburg and Christian 1933, p. 405)

Warburg also pressed much more specific objections to Wieland's model. As we have seen, he contended that the ability of hydrogen cyanide to inhibit cellular oxidation discredited Wieland's theory. He reasoned that, if the mechanism were hydrogen activation, there would be no process for hydrogen cyanide to inhibit.[19] Here Warburg's commitment to a single mechanism seems particularly evident.

Warburg's response to another objection brought by Wieland is also illustrative of his commitment to direct localization. Wieland's objection focused on the occurrence of auto-oxidizable reactions that did *not* involve heavy metals such as iron. Warburg countered by showing that many of these reactions did in fact involve minute quantities of iron or other heavy metals. Thus, cysteine appeared to be auto-oxidizable, but Warburg established that both hydrogen cyanide and pyrophosphate could inhibit the auto-oxidation of cysteine, a result difficult to explain on any other grounds than that a heavy-metal catalyst was involved. Warburg then developed techniques for removing traces of iron from cysteine compounds and showed that once cysteine was purified, it became less auto-oxidizable (1927; for additional examples, see Warburg 1949, ch. 5). Warburg was, thus, committed to showing that only one factor was causally responsible for respiration and to establishing this factor's identity.

4. CONCLUSION: DIRECT LOCALIZATION AND COMPETING MECHANISMS

In the previous chapter we emphasized the identification of a locus of control and the assumptions it involves. We have treated this as an application of two heuristics, decomposition and localization: we decompose the activities of nature and localize some of them in a particular system. Often a single system becomes a locus of control for a variety of activities, and the explanatory task becomes one of explaining this variety. Having identified a system as a locus of control, we treat it as a complex of components that are again subject to decomposition, and we localize systemic functions within components. It is then an empirical matter to inquire whether the behavior of the system as a whole can be explained in terms of the behavior of the identified component systems, and what kind of organization is needed to do so. The brief flowchart presented at the end of the previous chapter can thus be supplemented as in Figure 4.2.

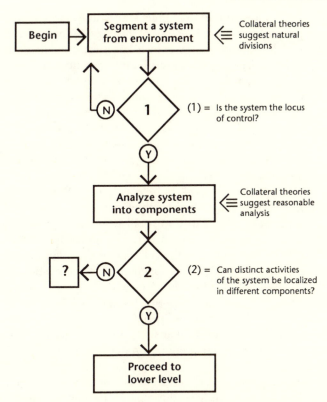

Figure 4.2. The Second Choice-Point. Having identified a system as a unit responsible for some function, one subsequent step involves decomposing the system, treating it as a complex of components, and localizing responsibility for systemic functions in specialized components. Localization and decomposition are often strongly influenced by collateral theories.

Direct localization depends critically on the ability to analyze the system into subsystems that operate in relative independence of one another. This analysis can be structural or functional. Gall and his followers employed a structural analysis of the brain and a functional decomposition of behavior: the mind was conceived as a system of discrete faculties; the brain as a system of discrete physical organs, each with a characteristic capacity. The analysis of the system need not be structural, though it is often implicitly supposed that it could be structurally analyzed even when the operative criterion is functional. Wieland and Warburg focused on identifying the agent of intracellular oxidation. Neither researcher had the physical information required to conduct a structural analysis into physical units or to provide a structural analysis of the enzymes involved. Warburg did postulate physical structures (which later were equated with mito-chondria) as the surfaces on which respiration was effected, but these did not play any significant role as his model unfolded and developed. War-burg and Wieland thus relied on a functional decomposition of the system.

The choice between structural and functional analysis was guided by available theoretical considerations—collateral theories—and practical feasibility. Gall thought that a functional decomposition of mental capaci-ties was afforded by his identification of different traits and capacities. The structural analysis into organs of the brain was provided by what ap-peared, at a morphological level, to be reasonably differentiated units of the brain. For Warburg and Wieland the functional subsystems were in-ferred from the way in which the overall respiratory process was con-ceived. Since hydrogen and oxygen did not interact under the general kinds of conditions prevailing in the cell, one or the other had to be acti-vated to facilitate the reaction. The functional characterization of the sys-tems was a natural outgrowth of the characterization of the problem, and this was a consequence of the theoretical framework in terms of which the problem of biological oxidation was posed.[20]

Given an analysis into constituent subsystems, the next problem is the identification of a likely component for producing a particular behavior of the system. If the system is analyzed into structural subsystems, then the natural problem is one of finding features of various structures that sug-gest that the structure will have the capacity in question. If the subsys-tems are differentiated functionally, then the functional decomposition will already have defined the subsystems in question, and the problem will be to isolate an appropriate structure. For Warburg and Wieland the relevant subsystems were already defined by the assumption that either hydrogen or oxygen had to be activated in order to react with the other. Whether one begins functionally or structurally the approaches are not incompatible, and, in general, the only satisfactory solution will be one that reflects both functional and structural properties.

Identifying an activity of a system with the activity of a component sub-system presupposes that the component structure will provide an ade-quate explanation of the system's capacities. Direct localization depends on demonstrating the *sufficiency* of the mechanisms thus isolated. We have noted a number of experimental strategies researchers deployed in garnering evidence relevant to this issue. The simplest cases rely on cor-relations between subsystems and the behavior of the system as a whole. According to Gall's view, the critical correlation was between the devel-opment of psychological faculties, as observed in behavior, and the size of the brain center in which the faculties are localized, as evidenced by cranial prominences. Positive evidence for direct localization was reflected in a high correlation. As we will see in Chapter 6, Broca subsequently introduced a novel and more compelling type of evidence for localization by correlating damage to a region with behavioral deficits. Given the assumption that faculties are localized in specialized regions, the correlation of deficits with behavioral irregularities provides strong evidence for particular localization claims. Broca had to rely on natural deficits, and that left open the question whether the correlation was accidental or was revealing of causal structure.

In the case of cellular respiration, researchers were able to introduce experimental techniques to interrupt or excite the localized center and thereby demonstrate the effects such changes had on system function. If an effect was still present after a component was removed, that provided strong evidence that the component did not play a critical role. Wieland's success in obtaining oxidations in the absence of oxygen afforded what he thought was compelling evidence that oxygen activation was not the critical component in oxidation. On the other hand, if removing a component did disrupt the function, that was good evidence that the component did play a central role. When Warburg found that cyanide inhibited cellular respiration, and he then provided reason to think that it acted by binding with cellular iron, he saw this as indicating that his iron-based oxygen-activating enzyme was critical to oxidation and cellular respiration. Excitation and inhibition can play important experimental roles by providing positive or negative evidence: Warburg found that adding iron salts to a cell facilitated oxidation, and he took the result to indicate that iron played a critical role in respiration; Wieland subsequently used the failure of ferric iron to induce catalysis as evidence against Warburg's model.

The success of direct localization will, of course, vary. In some cases the evidence may be compelling in favor of the view that a particular component is in fact responsible for the behavior of the system under study. Direct localization is not sufficient to describe a mechanism that can explain how a function is performed; it serves only to localize the "center" in which the function is performed, leaving the residual problem of explaining *how* the component performs the function in question. To explain subsystem operation, the next move would be to redirect inquiry to a yet lower level, possibly reapplying the kinds of strategies we have illustrated. We will discuss an example of this sort of theory development in Chapter 6.

It is important to see that the level of scientific debate is often over *which* component is responsible, and not over *whether* there is some isolated component. Wieland and Warburg were not arguing for decomposability or localization per se, but rather over the specific localization that was to prevail. Assuming there is a simple, direct localization is tantamount to assuming the system has a simple form of organization; it presupposes that there is relative independence of the component subsystems and correspondingly that the organization is either aggregative or composite. Even when this is correct, it may be true only to a first approximation. When the problem is one of explaining some particular dimension of a system's behavior, a localized component may explain variations only to a first-order approximation. Other components may turn out to be relevant. Once we have accounted for primary variations in the behavior of a system in terms of one component or a set of components, it is possible to focus on second-order interactions. In other cases, as with biological oxidation, there may be multiple components involved at the same level of organization. The task then becomes one of discovering the interactions between those components and showing how, together, they constitute an appropriate mechanism. The cases examined in Chapters 7 and 8 explore this strategy.

If we become convinced that no direct localization can be correct, even as a first approximation, and that we cannot attribute the behavior of a system to any one component, there are a variety of options. One is the path followed in the case of cell respiration. We may accept that more than one component is involved in producing the behavior and seek a systematic account of the components' interaction. In other cases, however, researchers have opted to reject decomposition and localization altogether, and consequently abandoned the search for mechanisms. It is to such cases that we turn in the next chapter.

CHAPTER 5

The Rejection of Mechanism

1. Introduction: Mechanism and Its Opponents

Not all scientific investigators see the development of mechanistic explanations as a critical constraint on their models. Even when there is general agreement in identifying a particular higher-level system as a locus of control for some phenomenon, those who do not accept a mechanistic paradigm may reject any attempt at further decomposition and localization of the sort described in Chapter 4. Decomposition and localization are seen as yielding only spurious explanations. As we will illustrate with several cases in this chapter, decomposition has been rejected by some for failing to provide an adequate explanation of life or of mind. Critics of decomposition and localization frequently attack the view in its simple form, what we have called direct localization. Whether or not these objections apply to more complex forms of localization such as those we describe in Part III, they do sometimes catch the shortcomings of direct localization. To mechanists it appears that those who repudiate localization are thereby repudiating scientific explanation, just as to opponents it appears clear that direct localization is a bogus explanation.

These debates are often driven by a variety of disparate concerns, from the religious and philosophical to the empirical and experimental.[1] Thus, in the evolving dispute between epigenesists and preformationists, Harvey and Descartes embraced the view that development must involve the emergence of organized forms from undifferentiated forms under the influence of the environment as a virtual corollary of their mechanistic commitments. In contrast, by the late seventeenth century, Malebranche, Malphigi, and Swammerdam began to look to development as the unfolding of preformed structures. Yet this shift to preformationism did not signal an abandonment of mechanism so much as a recognition of its limitations. The Cartesian vision of a mechanistic physiology was unable to explain the regular development of organisms, or the differentiation of organs and their reproduction. Others, such as Maupertuis, moved away from mechanism in a different fashion, allowing even the smallest elements in living organisms a capacity for intelligence and memory. Differences in the conceptions of acceptable explanations and acceptable methods loomed large in these debates, as in the other cases we will discuss.

Those who reject the search for mechanistic explanations have flown a variety of banners. We shall be concerned with only a few. *Dualism* (with regard to psychological activities) and *vitalism* (with regard to physiological activities) are two labels commonly adopted by opponents to mechanism—or applied to them by their mechanistic adversaries. The positive doctrines espoused by mechanists and antimechanists vary widely, as do the corresponding strengths and weaknesses. What they hold in common is hardly more than the view that something other than mechanistic processes are essential in understanding life and mind. Such a view can be held for quite compelling reasons and does not simply represent an irrational resistance to mechanism.

In the case of psychological theory, the most prominent opponents of mechanism are Cartesian dualists, though *epiphenomenalists* such as T. H. Huxley (1863, 1874), and *emergent materialists* such as R. W. Sperry (1969), count no less as opponents of mechanistic psychology. All, however, are mechanists in pursuing physiology, and all deny that the resources of physiology will suffice in assaulting the citadel of the mind. It is, in large part, their attempt to pursue mechanism that leads them to vitalism or dualism. Psychological dualists, such as Karl Popper and John Eccles (1977), are clear on what additional element they think necessary to deal with the phenomena of consciousness. With Descartes they maintain that cognition or consciousness depends on a different kind of substance, that the material substrate of even the most complex machines is inadequate to yield cognitive or conscious processes; moreover, the differences in substance foreshadow differences in capacities. In its most famous formulation, in the seventeenth century, this substance is not an alternative kind of matter out of which one might design engines of thought, but a substance whose primary attribute *is* thought. Dualists do not propose to explain consciousness or thought, but rather take them as primitive. Epiphenomenalists and emergent materialists eschew novel substances, but embrace the novel character of the conscious processes that pushes them beyond the reach of ordinary mechanistic explanations.

The vitalist position in physiology is even more varied. Some vitalists do hold out for the existence of a vital force or power and attribute to it the capacity to make ordinary matter into a living substance. But even here there is variety: Some vitalists treat this vital power as a unique entity in nature, while others, such as Justus Liebig, see it as quite closely connected with other, such as gravitational and mechanical, forces. Moreover, a significant number of vitalists are better thought of as emergent materialists or as *holists.* They do not posit a special substance or even acknowledge special powers, but argue that the distinctive properties of living organisms only appear once ordinary matter is organized in a particular way. They thus affirm a version of holism according to which the properties of life are treated as properties of the whole that cannot be refined into the properties of the parts, even when relations are taken into account. Some researchers, such as those physiologists Lenoir (1982) refers to as "teleomechanists," claim only that the special properties of life result from the way matter is organized in living things. This organization accounts for the teleological character of living systems; moreover, it could not be the result of simple underlying mechanical causes. For other scientists, such as those adopting some equipotentiality of particular systems, the holism is even more clearly an opposition to mechanism. The claim is made that because there is no differentiation of function between parts of the system, one cannot view the whole system as machine assembled from such parts.

In this chapter we shall examine four episodes in the life sciences involving researchers who did repudiate decomposition and localization, as well as the mechanistic assumptions on which these strategies rested, but did not thereby reject the importance of scientific investigation. These researchers included Pierre Flourens, Xavier Bichat, Theodor Schwann, and Louis Pasteur. The views expressed, at least in retrospect, were wrong, despite the eminence of the scientists involved. Our interest, however, is not in whether the views were erroneous, and even less in how the views were subsequently disproved, but in the factors that led them to repudiate mechanism and the effect this rejection had on the development of their disciplines. The terms in which their opposition was cast had implications both for the development of their own programs of research and for the development of those of their opponents. As always, the positive program of research is partly characterized and defined in contrast to the most significant alternative.

2. FLOURENS AND THE INTEGRITY OF THE NERVOUS SYSTEM

We noted in the last chapter that there were a variety of controversies, both in academic circles and in more popular forums, concerning phrenology. The first major theoretical challenge to Gall came from Pierre Flourens, with the support of Georges Cuvier. Flourens had flirted with phrenology early in his career, and even attended a course with Gall in 1815, but as he came increasingly under the patronage of Cuvier[2] he came to oppose the materialism he had once found attractive and became the premier defender of Cartesian dualism. Flourens's experimental results were heralded at the time as decisive refutations of specificity of brain action and were accepted doctrine for nearly forty years. Flourens opposed Gall for his materialism first and foremost, and secondly for his organology.[3]

Flourens isolated two propositions, which he took to be central to Gall's views: first, that the understanding resides exclusively in the brain; and second, that "each particular faculty of the understanding is provided in the brain with an organ proper to itself" (1846, p. 18). There is, he said, "certainly nothing new in the first one, and perhaps nothing true in the second one" (ibid.). The second proposition became the immediate target of Flourens's famous and elegant experiments supporting the equipotentiality of the nervous system. In these experiments Flourens extirpated parts of the cerebral lobe of pigeons. He found that the animal's visual abilities systematically weakened with successive excisions until they were totally lost. At the moment that sight was lost, Flourens found that other sensory faculties were lost as well. The conclusion was expressed in the principle of the "Unity of the Nervous System":

> Independently of the *proper action* of each part, each part has a *common action* with all the others, as have all the others with it. (1824, p. 241)

This is a straightforward denial of decomposability and an insistence on the integration of neural functioning. The result was a defense of an equipotentiality of action within the cerebral hemispheres:

> All sensations, all perceptions, and all volition occupy concurrently the same seat in these organs. The faculty of sensation, perception, and volition is then essentially one faculty. (Ibid.)

Flourens did not categorically reject faculties of the mind; however, though granting that the faculties were nominally distinct, he said that if a commitment to the independence of the faculties also meant that "each faculty is a real understanding"—as it manifestly did, in the case of Gall and Spurzheim—then it would destroy the "unity of the understanding." This view lay at the heart of Flourens's metaphysics and was supposedly vindicated by his experimental results:

> It has been shown by my late experiments, that we may cut away, either in front, or behind, or above, or on one side, a very considerable slice of the hemisphere of the brain, without destroying the intelligence. Hence it appears, that a quite restricted portion of the hemispheres may suffice for the purposes of intellection in an animal.
>
> On the other hand, in proportion as these reductions by slicing away the hemispheres are continued the intelligence becomes enfeebled, and grows gradually less; and certain limits being passed, is wholly extinguished. Hence it appears, that the cerebral hemispheres concur, by their whole mass, in the full and entire exercise of the intelligence. (1846, p. 34)

According to this view, the brain became merely the organ on which the immaterial mind worked its will, and the various mental activities became but aspects of a single and unitary mind. The mind, Flourens says, citing Descartes, is indivisible and lacking parts.

Flourens did not wholly deny differentiation in brain structures. Indeed, he is often regarded as the first to demonstrate experimentally that distinctive structures have their own distinctive functions. Flourens noted that destruction of specific regions of the brain leads in turn to specific deficits: removal of the cerebellum leads to loss of locomotive action; removal of other regions affects sight only; other regions were specialized for respiration; and removal of the hemispheres leads to a loss of what Flourens called "understanding." Moreover, these losses left the other capacities intact. Flourens thus concluded that there are *exactly* four "particular organs":

1. The cerebellum, regulating locomotion.
2. The tubercula quadrigemina, regulating sight.
3. The medulla oblongata, regulating respiration.
4. The "brain proper."

The latter was "the seat, and the exclusive seat of intelligence" (ibid., p. 31). He concluded:

> Everything concurs then to prove, that the encephalon, in mass, is a multiple organ with multiple functions, consisting of different parts, of which some are destined to subserve the locomotive motions, others the motions of respiration, &c., while *one single one, the brain proper, is designed for the purposes of the intellection.*
>
> This being conceded, it is evident that the entire brain cannot be divided, as the phrenologists divide it, into a number of small organs, each of which is the seat of a distinct intellectual faculty; for the entire brain does not serve the purposes of what is called the intelligence. (Ibid., pp. 32–33)

The unity of the mind came to be reflected in the unified and coordinated activity of the cerebral hemispheres, or in their "common action." The higher cognitive functions assigned to the understanding could not be further decomposed. Thus, having assigned different functions to the four regions of the brain, decomposition and localization could do no more; the brain proper could not be divided, so its higher cognitive functions could not be localized in its components.

Accordingly, Flourens regarded Gall's views as issuing from a false abstraction. Flourens observed that Gall thought "each faculty of the soul must have its proper organ; in one word, he looks upon the outer man, and constructs the inner man after the image of the outer man." Gall thus began by observing a diversity in the expressions of intelligence and projected a diverse array of mechanisms. Gall's whole philosophy, according to Flourens, "consists wholly in the substitution of multiplicity for unity"

(1846, p. 47); Gall substituted "a multitude of little understandings or faculties, distinct and isolate" for "General Intelligence," or what was properly the "Understanding" and was correctly understood as a "unit faculty" with a multiplicity of applications. Flourens then concluded "that an explanation, which is words merely, adapts itself to any and to every thing" (ibid., p. 38):

> You observe such or such a penchant in an animal, such or such a taste or talent in a man; presto a particular faculty is produced for each one of these peculiarities, and you suppose the whole matter to be settled. You deceive yourself; your faculty is only a word—it is the name of the fact—and all the difficulty remains just where it was before. (Ibid., p.39)

The challenge is a familiar one to the contemporary ear, and was hardly new with Flourens. Opponents of faculty psychology, from Galen and Descartes to Herbart and Moliére, traditionally objected that explanations in terms of faculties were spurious precisely because they simultaneously classified behavior and explained it *in the same terms*; that is, behavior was classified, and the classification was projected to a lower level, which then supposedly explained the behavior. Flourens likewise regarded the "explanations" of the phrenologists as nothing more than redescriptions of the very phenomena needing to be explained. Such explanations were considered either vacuous or speculative. In either case, they were to be rejected.

The response from supporters of faculty psychology was equally familiar. Gall rejoined that the "intellectual faculty and all its subdivisions, such as perception, recollection, memory, judgment, and imagination, are not fundamental faculties, but merely their general attributes" (Gall and Spurzheim 1810–1819, 4:327). His mechanistic commitments made it impossible to understand how "there should be any peculiar organ of the reason" (ibid., 4:341). There was no hope of a mechanistic explanation of reason understood as a general, all-purpose capacity. The only hope for mechanism was a decomposition of the functions of the mind. Flourens also saw decomposition and mechanism as integral to one another; accordingly, he denied that either was viable. What supposedly showed the reality of the separate faculties, according to Gall, was the possibility of their being independently manifested. If they could be separately manifested they must be distinct, even if they were generally expressed together. The fact that they could be simultaneously affected by extirpation or stimulation, as Flourens had shown, could do nothing to show that they were one. What was required, rather, was a demonstration that they could *not* be differentially extirpated.

In addition to methodological criticisms of Flourens, the phrenologists raised technical objections to the use of artificial lesion data:

Several natural philosophers have endeavored by mutilations, viz. by cutting away various parts of the brain, to discover their functions. These means have been pursued without fruit and will remain useless. They are too violent, and several faculties might be retained without being manifested; at all events they cannot teach more than may be ascertained in the healthy state. . . .

The best method of determining the nature of the cerebral functions, is that employed by Phrenologists: it is to observe the size of the cerebral parts in relation to particular mental manifestation, and it is the third principle of phrenology, that in the same individual, larger organs show greater, and smaller organs less energy. (Spurzheim 1832, p. 12)

Gall observed, in an extended discussion occupying much of volume 6 of *On the Functions of the Brain and Each of Its Parts*, that ablation could not provide a useful experimental method because it required researchers to know beforehand the extent of an organ in order to get significant results. One would need more precise methods than were technically feasible and more adequate controls.[4]

The conflict remained at an impasse. Phrenological evidence was insufficient to show that decomposition and localization did any serious explanatory work. The antimechanistic stance of Flourens and his supporters denied the relevance of the data, treating it as a simple artifact. For some time Flourens's position was the dominant establishment view. It would not be until Broca (see Chapter 6) produced his own evidence concerning localization of function that this orthodox conception of brain action would be successfully challenged.

3. THE VITALIST OPPOSITION TO MECHANISTIC PHYSIOLOGY

As we noted at the beginning of the chapter, vitalism embraces a variety of positions, ranging from the staunchly antimechanistic to those generally congenial to mechanism, but maintained that it was necessary to expand the range of forces in order to explain processes found in living beings. We will restrict our focus in this section to two positions to which the term *vitalism* has been attached: Bichat's rejection of mechanistic accounts of respiration, and Schwann and Pasteur's rejection of chemical accounts of fermentation. These views, like those of Flourens', are of interest in large part because they represent alternatives to mechanistic programs examined elsewhere in this book. They are also interesting because these researchers understood the mechanistic program—indeed, in some domains were advocates of it—but saw compelling reasons to insist on limits to that program.

Bichat's Opposition to Lavoisier's New Chemical Physiology

Lavoisier, in addition to contributing to a revolution in chemistry, applied the new chemical theories to physiology, proposing that biological respiration was comparable to ordinary combustion, only involving a fire without a flame. Numerous researchers, many with roots in chemistry, pursued the path proposed by Lavoisier and attempted to explain a number of physiological functions in purely chemical terms. (In Chapter 3 we briefly discussed attempts to develop evidence for various possible sites for animal respiration.) In addition to Lavoisier, Bertholet and Gay Lussac set out to analyze the chemical constitution of plants and animals, generally in terms of their elementary composition, and to explain the processes by which food substances became transformed into the tissue of an organism (see Holmes 1974). To other investigators, however, the attempt to explain the processes of life in chemical terms was misguided. They were convinced that processes occurring within living organisms were different in kind from those occurring outside living organisms.

One spokesperson for this opposition was Xavier Bichat, a French anatomist and surgeon toward the end of the eighteenth century (see Albury 1977). In some respects Bichat seems to be a paradigmatic mechanist, explaining physiological functions by tracing them back to the properties of components localized within the system. He asked what physical structures in the body could account for the variety of phenomena exhibited by the various organs of plants and animals. He differentiated several kinds of tissues in the body, each with its own primitive properties. Bichat here carried out a program of direct localization: Having differentiated twenty one types of tissue, he attributed a different and distinctive set of traits to each. He explained that these traits appeared in the organs themselves because they were composed from tissues that already possessed the traits. Various organs differed in their properties because they were composed of different tissues with different intrinsic qualities. The properties of the organs were thus "explained" by showing that the organs contained tissues that themselves exhibited the properties in question. Bichat compared this explanatory strategy to decomposing chemical compounds into chemical elements:

> Now these separate machines [the organs] are themselves formed by many textures [tissues] of a very different nature, and which really compose the elements of these organs. Chemistry has its simple bodies, which form, by the combinations of which they are susceptible, the compound bodies; such are caloric, light, hydrogen, oxygen, carbon, azote, phosphorus, &c. In the same way anatomy has its simple textures, which, by their combinations four with four, six with six, eight with eight, &c. make the organs. (Bichat, 1801/1822, cited in Hall 1951, p. 69)

Bichat thus seemed to be well embarked on a typical mechanist program, pursuing a strategy of direct localization. The next step in the mechanist program would have been to transfer the inquiry to a lower level, decompose the functions assigned to the tissues, and localize these functions within components of the tissues themselves. One would thus explain how the tissues had the properties they did in terms of their composition. The following move would have been to frame this model in terms of the kind of chemical theory being developed by Lavoisier and other chemists. Yet, at this juncture, Bichat rejected the mechanistic program and advanced two arguments against any attempt to account for the properties of tissues in more basic terms. One was based on a fundamental indeterminacy in vital processes; the other depended on their self-regulation.

In characterizing the properties of tissues, Bichat used two concepts central in the investigations of Albrecht von Haller: *irritability* (or *contractility*) and *sensibility*. Irritability, by Haller's definition, was the inherent ability of muscles to contract. Sensibility was an attribute of the nerves that allowed response to external stimuli. Haller treated these as basic features of living organisms and argued that although these activities could be measured, they could not be explained in more fundamental terms. Moreover, irritability was an attribute of animal muscle-tissue that was independent of the soul. Haller remarked, in an early passage, that the heart of dead animals continued beating, from some "unknown cause . . . hidden in the fabric of the heart itself" (1739, 2:129). The ultimate source of these attributes was God, who "gave to bodies an attractive force and other forces, which once received are exercised" (1756–1760, 1:xii).

According to Bichat the problem was even more serious than Haller had envisioned. Bichat argued that one could not even obtain accurate measurements of these basic properties because the sensibility and irritability exhibited by the tissues continually varied and so could not be determinately measured. This indeterminacy, for Bichat, marked a critical difference between the organic and inorganic worlds. While nonliving entities adhered to strict deterministic laws, living entities did not. Living things were at every instant undergoing some change in degree and kind:

> They are scarcely ever the same. . . . In their phenomenon nothing can be foreseen, foretold nor calculated; we judge only of them by their analogies, and these are in the vast proportion of instances extremely uncertain. . . . To apply the science of natural philosophy to physiology would be to explain the phenomena of living bodies by the laws of inert body. Here . . . is a false principle. (Bichat 1801/1822, p. xx; cited in Goodfield 1960, p. 69)

On account of this indeterminacy, Bichat objected to applying the principle of deterministic science to living systems:

One calculates the return of a comet, the speed of a projectile; but to calculate with Borelli the strength of a muscle, with Keill the speed of blood, with Lavoisier the quantity of air entering the lungs, is to build on shifting sand an edifice solid itself, but which soon falls for lack of an assured base. This instability of vital forces marks all vital phenomena with an irregularity which distinguishes them from physical phenomena [which are] remarkable for their uniformity. It is easy to see that the science of organized bodies should be treated in a matter quite different from those which have unorganized bodies for their object. (1805, p. 81; cited in Goodfield 1960, p. 68)

While Bichat's first argument focused on indeterminacy, his second argument contended that living things evaded and opposed the laws of nature. This provided them with self-regulatory capacities, which, as Bichat saw, were necessary if living organisms were to survive in the face of natural forces that would tend to destroy them. For example, the temperature of ordinary objects gradually comes into equilibrium with the temperature in the environment. However, warm-blooded animals will oppose this tendency and maintain themselves at a set temperature regardless of their environment. Thus, Bichat came to conceive of the vital properties of living organisms as standing in opposition to those of inorganic nature. The properties of inorganic nature tended to tear down living organisms, and these organisms' vital properties were required to oppose these inorganic properties. He consequently characterized life as "the sum of all those forces which resist death" (1805, p. 1). Since living organisms opposed the natural tendency of nonliving nature, they could not be explained in terms of the processes of nonliving nature alone.

Although Bichat thought the indeterminacy and self-regulation of living organisms ruled out any mechanistic attempt to explain the properties of tissues in more basic terms, he did not therefore oppose empirical and experimental inquiry into the operation of living organisms. His tissue theory was the result of just such an inquiry, and in it he developed experimental techniques to identify the particular forms of sensitivity and irritability manifested by different tissues. What Bichat's principles did, though, was limit the inquiry into these tissues to a descriptive mode: empirical investigation should only *describe* what sorts of sensibility and irritability different tissues exhibited and not try to explain *why* the tissues manifested these properties by appealing to other properties found in nonliving matter.[5] Thus, Bichat, like Haller, treated these forces as basic forces in nature: "To create the universe God endowed matter with gravity, elasticity, affinity, etc., and furthermore one portion received as its share sensibility and contractility" (ibid.).

Bichat rejected the mechanistic program at the point that requires an explanation of the properties of tissues in terms of properties of inanimate matter, because these tissues seemed to possess features that were not exhibited by, and in fact seemed opposed to those of, inanimate matter. It appeared to him quite inconceivable that any purely physical machine could possess these properties. Indeed, if direct localization is the model of how to develop a mechanistic explanation, his objections appear well taken; for direct localization assumes that we will find some lower-level component with the properties of tissues, and it is clear that no purely chemical constituent could have such properties. Bichat was quite aware of the attempts by his contemporaries to carry out mechanistic investigations, but he contended that their efforts would necessarily be futile and that sensibility and irritability must be accepted as basic properties. One could characterize them and differentiate tissues in terms of their possession of these properties, but one could not explain these properties mechanistically. Proper scientific explanation was limited to this descriptive mode, just as Newtonian theory was restricted to describing the phenomena. And just as Newton had to accept gravity as a fundamental force, irreducible to mechanical push and pull, the physiologist had to accept sensibility and irritability as fundamental forces, irreducible to mechanistic chemistry. Vital forces could not be observed and were therefore not proper objects for mechanistic explanation.

François Magendie, one of the most prominent physiologists in the generation after Bichat, as well as others such as Johannes Müller, repudiated Bichat's case for indeterminism and brought a greater quantitative focus to physiological research. Yet Magendie remained a vitalist and rejected attempts to provide a mechanistic explanation of all living phenomena. While Bichat had drawn a parallel between vital forces and Newtonian forces (like gravity), all the while seeking to develop a physiological description that accepted indeterminacy, Magendie insisted that without precise quantitative laws to describe these forces, no science was possible. Magendie (1809) also rejected the idea of developing a taxonomy of tissue types in favor of examining organs and their interactions within animals. As he expressed it, his aim was "to abolish the two vital properties known under the names of animal sensibility and animal contractility and to consider them as functions." To do this Magendie popularized the technique of vivisection, whereby one could interrupt the activity of a component within the system and see what consequences it had on the overall functioning of the system. He was able to differentiate the sensory function of the posterior roots of the medulla and the motor function of the anterior roots, as well as to identify the control of the fifth cranial nerve over touch and of the cerebellum over balance (Magendie 1821; see Albury 1977).

Like Bichat, Magendie seemed well on the way to a decomposition of vital functions and a localization within bodily tissues. While Magendie did contribute information that was critical to developing a mechanistic explanation of vital functions, he did not embrace mechanism applied to these functions. This is not to deny that he attempted to unravel the chemical and physical processes operating within the organism. For example, he conducted various studies on the effects of poisons in the body. While part of his goal was to isolate the active agent in the poisons, he did *not* try to analyze everything at this chemical level. He performed physiological experiments to determine where poison entered the body: by separating the rear quarter of a dog so that only the blood vessels remained intact and then showing that poison injected into that portion still entered the animal's circulation and took effect (cf. Olmstead 1944), he established that poison did not enter only through the lymphatic vessels. Yet, in carrying out these physiological experiments, Magendie accepted that there was a level of analysis beyond which he could not go. He was unable to explain nutrition or action in mechanical terms, as these processes took place at a level that was imperceptible and hence not subject to experimentation, at least given the tools at hand. His opposition therefore appears more pragmatic than principled: without appropriate experimental techniques to identify components and their contributions, he saw there was no hope of discovering the operative mechanisms.

Bichat's and Magendie's opposition to the mechanistic explanation of physiological processes in terms of chemical mechanisms provides an interesting study in contrasting styles. Bichat pressed principled arguments designed to show that physiological processes exhibited distinctive properties, inexplicable in physical or chemical terms, which therefore could not be identified with physical or chemical processes. Magendie, by contrast, offered technical rather than principled arguments against mechanism in physiology: lacking the means to reveal physical or chemical processes underlying vital processes, any proposed mechanistic account was doomed to be idle speculation.

The Explanation of Fermentation

Bichat's objections to mechanism were general and sweeping; he opposed *any* attempts to account for the properties of living tissues in mechanistic terms. By the middle of the nineteenth century, however, a more focused opposition arose, one directed at fermentation. Proponents of this objection tried to show that fermentation was one physiological process irreducibly linked to living cells.[6]

The framework for modern chemical thinking about fermentation stems from Lavoisier, who viewed it as a two-part process: sugar was broken down into simpler components, one of which was subsequently oxidized:

> The effects of venous fermentation upon sugar is [*sic*] thus reduced to the mere separation of its elements into two portions; one part is oxygenated at the ex-

pense of the other, so as to form carbonic acid, while the other part, being disoxygenated in favor of the former, is converted into the combustible substance called alkohol; therefore, if it were possible to re-unite alkohol and carbonic acid together, we ought to form sugar. (Lavoisier 1799, p. 196)

At the beginning of the nineteenth century, Gay-Lussac (1810) and Thénard (1803) extended the inquiry of Lavoisier, establishing (in modern symbolism) the following general equation for fermentation:

$$C_6H_{12}O_6 \Rightarrow 2CO_2 + 2C_2H_5OH.$$

The only thing that seemed to require explanation was what prompted the decomposition of sugar into these components. Gay-Lussac's proposal typified early-nineteenth-century thinking. Following Appert's (1810) discovery that one could preserve food by sealing it from air and heating it, Gay-Lussac proposed that exposure to ordinary oxygen was what initiated the decomposition process (see Fruton 1972).

The case for the involvement of living organisms in fermentation was developed by Cagniard-Latour (1838), Schwann (1837), and Kützing (1837).[7] All three were microscopists. Cagniard-Latour's first observations of yeast, made at the outset of the nineteenth century, showed yeast to be a fine, granular material. New microscopes, introduced in the 1830s, allowed him to observe yeast reproducing through budding, an activity typical only of living organisms. He also ascertained that the number of globules was always proportional to their weight. Thus, he concluded that yeast could not be a simple chemical substance, but was instead a living body: "brewer's yeast, this widely-used ferment, is a mass of small globular bodies that can reproduce themselves . . . and not simply an organic or chemical substance, as has been supposed" (Cagniard-Latour 1838, p. 221). Because it could not move, Cagniard-Latour concluded that yeast was a plant. He further contended that its production of alcohol and carbon dioxide from sugar was the result of its growth.

Schwann approached the identity of yeast from a different direction. He had been addressing the question of spontaneous generation and also followed up on Appert's discovery that one could suppress fermentation and putrefaction by heating and sealing organic material from air. Schwann showed that sealing off air was unnecessary to stop putrefaction as long as all air reaching the organic material was heated. He interpreted these results as showing that putrefaction was due to living organisms, which were present in ordinary, but not in heated, air. These organisms fed on the dead organic matter. Since putrefaction and fermentation were regarded as very similar processes, he proposed extending these findings to fermentation. He discovered, however, that normal yeast cells could maintain fermentation even when provided with only heated air, but that if they were boiled, fermentation could not continue. This led him to propose that yeast is a living organism, a fact further supported by his observation that corpuscles in brewer's yeast multiplied during the course of fermentation and that arsenic halted fermentation. He found additional support in the fact that the rate of fermentation increased in the presence of nitrogenous substances, which he attributed to the fact that yeast needed nitrogen for normal growth. Like Cagniard-Latour, Schwann interpreted alcoholic fermentation as a byproduct of the normal nutritive process in yeast:

> Alcoholic fermentation must therefore be regarded as a decomposition effected by the sugar-fungus, which extracts from the sugar and a nitrogenous substance the materials necessary for its own nutrition and growth, and whereby such elements of these substances (probably among others) as are not taken up by the plant preferentially unite to form alcohol. (1837, p. 192)

The proposal that fermentation required the activity of a living organism was vigorously attacked by a number of researchers who remained convinced that it was a purely chemical reaction. One of the strongest proponents of the chemical view was Jacob Berzelius. In his paper introducing the concept of a *catalyst*, he had proposed that fermentation was a catalytic process like others he considered:

> Of all the known reactions in the organic sphere, there is none to which the reaction bears a more striking resemblance than the decomposition of hydrogen peroxide under the influence of platinum, silver, or fibrin, and it would be quite natural to suppose a similar action in the case of the ferment. (1836; cited in Keilin 1966, p. 20)

Berzelius (1839) also criticized the proposals of Cagniard-Latour, Schwann, and Kutzing, arguing that yeast is simply a chemical substance, not a living organism. He proposed that it participates in fermentation and forms a precipitate resembling living cells.

The conflict over whether yeast cells were living organisms was bitter and heated. In 1838 Turpin prepared a paper for the French Academy in which he defended Cagniard-Latour's views. When a description and partial translation of the paper appeared in the *Annalen der Chemie* (Turpin 1839), it was followed by an anonymous article, widely attributed to Wohler and Liebig (the editors of the journal). The article (Anonymous 1839), entitled "The Unravelled Mystery of Spirituous Fermentation," ridiculed the view of yeast cells as living organisms by treating the view as holding that yeast cells were little animals shaped like distilling glasses, without teeth or eyes but with a stomach, an intestine, and urinary organs. According to the article, yeast cells digest sugar in their stomach, producing carbon dioxide and alcohol, the latter being expelled through a bladder the shape of a champaign bottle (see Fruton 1972, pp. 50ff.).

Although satirical, the article did raise a legitimate objection: identifying fermentation with the life processes of an organism did not itself produce any explanatory gain. Attributing fermentation to the digestive processes of a living organism did not explain the process unless one could detail the chemical reactions through which the digestion occurred. On the other hand, these authors overreacted to the idea that living organisms might figure in the process, thinking it was incompatible with any form of mechanistic explanation of the process. Schwann had already pointed to a way in which linking fermentation with life could still be compatible with a restricted form of mechanistic explanation—one construing the components of the cell as ordinary physicochemical constituents that are transformed by being adsorbed into living cells. While Schwann was not the first to have observed cells or to have claimed that they were the basic structural unit of living organisms, it was through his work that the cell theory gained general acceptance. The reasons for this were complex. One is that, due to improvements in microscope design and Schwann's ability as a microscopist, his observations were more credible than those of his predecessors. A more important factor, however, had to do with his theoretical argument that all cells, which may appear quite different in different tissues, are really the same kind of morphological unit, because they are formed in the same manner. He presented microscopic evidence to support the claim that cells are formed by a process analogous to crystal formation. Despite morphological differences that might appear later in development, the common origin of the various cells showed them to be the same kinds of living units. This was a highly mechanistic proposal, and the fact that he was advancing such a model encouraged a general acceptance of his cell theory (see Bechtel 1984a). Other researchers, including von Mohl, reported observing cell division, but could not offer a mechanical model to account for the process.

Schwann's commitment to mechanism may seem at odds with his defense of the claim that living cells, and not simple chemical structures, are the agents responsible for fermentation. Schwann, however, did not conceive of himself as totally giving up mechanism in defending his account of fermentation. He viewed living systems as themselves mechanical systems, but thought that the special kinds of structures found in living systems were the particular mechanical systems required to carry out tasks like fermentation. This becomes clear in the third part of his *Microskopische Untersuchungen* in which Schwann develops what he considers to be his speculative—and mechanistic—"theory of the cell." He divides the functions of cells into two kinds. He calls the first "plastic phenomena"; these relate to the process by which cells develop. The second he calls "metabolic phenomena"; these "result from chemical changes either in the component particles of the cell itself, or in the surrounding cytoblastema."

It is clear that Schwann intended metabolic phenomena to include such processes as fermentation and respiration, and that he expected to explain these phenomena in terms of the specialized constitution of the cell:

> The cytoblastema [i.e., the intercellular fluid], in which the cells are formed, contains the elements of the materials of which the cell is composed, but in other combinations; it is not a mere solution of cell-material but it contains only certain organic substances in solution. The cells, therefore, not only attract materials from out of the cytoblastema, but they must have the faculty of producing chemical changes in its constituent particles. Besides which, all the parts of the cell may be chemically altered during the process of its vegetation. The unknown cause of all these phenomena, which we comprise under the term metabolic phenomena of the cell, we will denominate the *metabolic power*. The next point which can be proved is that this power is an attribute of the cells themselves, and that the cytoblastema is passive under it. We may mention venous fermentation as an instance of this. A decoction of malt will remain for a long time unchanged; but as soon as some yeast is added to it, which consists partly of entire fungi and partly of a number of single cells, the chemical change immediately ensues. Here the decoction of malt is the cytoblastema; the cells clearly exhibit activity, the cytoblastema, in this instance even a boiled fluid, being quite passive during the change. (1839, p. 197–98)

Schwann, thus, did not fully abdicate mechanism. He thought that the fermentative power of cells was due to their constitution, yet he insisted that this constitution is peculiar to living cells. The metabolic power associated with the living cells alters the passive cytoblastema, making it into cellular substance, and gives it the power to produce chemical reactions such as fermentation. To the degree that Schwann rejects mechanism, it is because he does not think that ordinary chemistry alone can explain fermentation. To do that, it is necessary to consider the metabolic power that organizes these constituents into the cell.

The controversy between those who took fermentation to be a purely chemical process and those who saw it as the activity of living yeast continued for several decades. In these discussions the concept of fermentation was applied far more generally than just to alcoholic fermentation. It was applied to any reaction where nitrogen-based organic material, present in small quantities, catalyzed a chemical breakdown of substances into simpler components, without being consumed in the reaction. Thus, it included the breakdown of sugar into lactic acid in the souring of milk, or that of sugar into butyric acid in the souring of bread; the transformation of gums, starch, dextrin, and mannitol into dextrose; the formation of ammonia from urea; and the transformation of proteins, starches, and fats into simpler products by saliva, gastric juice, or pancreatic juice. In many of these cases there was no indication of the involvement of living organisms, thus creating a tension between two types of accounts for different fermentations. The proponents of the chemical accounts argued that all fermentations ought to be simple chemical reactions, and they viewed those who advocated a special role for living organisms in some fermentations as engaging in special pleading. The chief combatants advancing these two alternatives were Justus Liebig and Louis Pasteur.

Liebig, after initially simply endorsing Berzelius's theory, later advanced his own, one more reminiscent of Stahl than of Berzelius. A ferment, according to Liebig, was nothing but an unstable organic product, and fermentation was a chemical reaction, a decomposition conceived of as akin to oxidation. He attributed the appearance of living globules to "gluten or albumin [taking] a geometric form which [became] altered during the process of fermentation of beer, wort, or vegetable juices" (Liebig 1839). His own theory attempted to integrate the observation that nitrogenous material undergoes spontaneous putrefaction, while non-nitrogenous material must be brought into contact with putrefying substances (such substances caused sugar to ferment rapidly when the two were brought into contact). Yeast simply accomplished this to a higher degree. Liebig attributed this power to the kind of chemical activity occurring in the nitrogenous substances and held that these substances were, in turn, capable of passing this activity to others. Thus, it was not a particular material, such as Berzelius's catalysts, that induced fermentation, but simply unstable substances capable of passing movement onto other substances. In the fermentation of beer, yeast was essential not because of any organic activity, but because, in dying, it released chemicals into the medium, and these chemicals in turn initiated fermentation. Liebig's theory of fermentation, along with various theories of animal and plant metabolism that he developed in subsequent years, held considerable influence among chemists for some time.

Pasteur defended the position of Cagniard-Latour and Schwann, holding that living cells were essential to fermentation. As he said, "organization and life" are the cause of fermentation. He was led, however, to take up the question of fermentation as a result of his concern with a quite different set of problems. Prior to attacking the question of fermentation, Pasteur had been a pure chemist little interested in problems relating to physiology. He apparently first considered fermentation and observed yeast under a microscope in 1856, when he was asked to consult on some problems facing the local brewing industry (see Connant 1948). His intellectual interest in fermentation, though, stemmed from a purely chemical problem—that concerning the ability of certain solutions to polarize light. Pasteur thought that this ability was correlated with the asymmetry of molecular structure, but this correlation seemed to fail with amyl alcohols, a by-product of fermentation; he then distinguished two amyl alcohol isomers, only one of which polarized light. Pasteur was further interested in how the capacity to polarize light arose, since only substances produced in living organisms seemed to possess this capacity.

The first paper to result from his research on fermentation (Pasteur 1858a) dealt not with alcoholic fermentation but with lactic acid fermentation, a reaction most readily observed in the souring of milk. In the mid-nineteenth century the two reactions were conceived to be quite different physiological processes for metabolizing glucose.[8] Lactic acid fermentation, in fact, seemed to offer strong support for Liebig's conception of fermentation reactions, as it did not appear to involve living organisms in any way (Gerhardt 1856). The common procedure for trying to eliminate microorganisms from biological material—boiling the substrate and heating the air—did not impede lactic fermentation. Pasteur's goal was to demonstrate that, despite appearances, a "new yeast" figured in lactic acid fermentation:

> Just as an alcoholic ferment exists, namely, brewer's yeast, which is found wherever sugar breaks down into alcohol and carbonic acid, so too there is a special ferment, a lactic yeast, always present when sugar becomes lactic acid, and that if any nitrogenous plastic material can transform sugar into this acid it is because it is a food suitable to the development of this ferment. (1858a, p. 28)

He prepared a culture of yeast, boiled it to kill the cells, and seeded it with a lactic ferment. After incubating for a day, this lactic ferment induced an active lactic fermentation in the new solution and a release of CO_2. Though Pasteur initially used brewer's yeast to make the culture, he was quick to point out that any "nitrogenous" substance (that is, any protein) would suffice. Using microscopic observations, he also set out to show that the new yeast was very similar in appearance to brewer's yeast, which led him to the thought that they might be "neighboring species."

In subsequent work Pasteur turned directly to the study of alcoholic fermentation and set out to refute Liebig's interpretation of the process. He showed that when yeast was added to a liquid medium of sugar, ammonium tartrate, and inorganic phosphate in which there was no substance that could be construed as spontaneously decomposing (as Liebig's model required), fermentation still ensued. This inducement of fermentation could only be attributed to yeast. Moreover, he showed that when fermentation occurred some of the mass of the sugar substrate was not accounted for in the output, which pointed to the fact that some of the sugar had been taken up by the yeast. From such evidence Pasteur concluded that fermentation was essentially a process associated with life:

> Alcoholic fermentation is a process associated with the life and the organization of yeast cells and not with their death or putrefaction any more than it is a

phenomenon of contact, in which case the transformation of sugar would be accomplished in the presence of the ferment without yielding anything up to it or taking anything from it. The chemical process of fermentation is essentially a phenomenon associated with vital action, commencing and ceasing with it. It is invariably accompanied by the growth and multiplication of yeast cells. (1860a, p. 359)

As to how yeast accomplished fermentation, Pasteur was agnostic. He allowed that the participation might come close to that envisaged in Liebig and Wohler's satire, but he also acknowledged the possibility that it might in fact be due to a peculiar chemical agent produced by the yeast that functioned as a catalyst:

Will one say that the yeast nourishes itself with sugar so as to excrete it in the form of alcohol and carbonic acid? Will one say on the contrary that the yeast produces, during its development, a substance such as pepsin, which acts on the sugar and disappears when that is exhausted, since one finds no such substance in the liquids? I have no reply on the subjects of these hypotheses. I do not accept them or reject them, and wish to constrain myself always not to go beyond the facts. And the facts only tell me that all the fermentations properly designated as such are correlative with physiological phenomena. (Ibid., pp. 360)

Pasteur thus did not explicitly rule out the possibility that fermentation could eventually be explained in chemical terms. He insisted that the process was connected with the life of a cell and could not be totally separated from it, and the possibility of a chemical explanation was set aside as hypothetical and not needing an answer. His own methodology required that he not "go beyond the facts." Insofar as he resisted the attempt to decompose the living system and to attribute fermentation to component parts, and instead insisted on correlating it with living cells, though, he was rejecting the development of a mechanistic approach to explaining fermentation. He abjured speculation.

Although Pasteur rejected any mechanistic research into fermentation, he did not reject the attempt to study fermentation empirically. In fact, in his subsequent research, he established one of the important features of yeast life, which is now known as the "Pasteur Effect": yeast develops and causes fermentation in the absence of oxygen, losing this ability when oxygen is present. Pasteur described this as a process of *vie sans air*, arguing that the yeast cells carried out metabolic activities very similar to those found in aerobic organisms, but that they drew their oxygen from the metabolites into which oxygen was unstably bound.

Liebig eventually came to accept the idea of yeast cells as living organisms, but he still did *not* accept fermentation as a physiological process of living organisms. He argued that fermentation resulted from the disintegration of yeast—that when yeast cells ceased to grow, the bond that held them together deteriorated and a motion resulted, which was then imparted to the sugar. Liebig argued that the contribution of the living organism was simply to generate the substance that subsequently induced fermentation:

> It is possible that the only correlation between the physiological process and the phenomenon of fermentation is the production, in the living cell, of a substance which, by some specific effect analogous to that which emulsin exerts on salicin and amygdalin, brings about the decomposition of yeast and of other organic molecules. According to this view, the physiological process would be necessary for the production of this substance, but would have nothing to do with the fermentation. (1870, p. 6)

Pasteur (1872) responded in a brief note in which he pointed out again that even when there were no substances to decompose in the manner Liebig proposed, yeast still induced fermentation.

Opposition to Pasteur also came from chemists who viewed his model as vitalistic. This is particularly true of Berthelot, who remarked, "As to the vitalist opinions adopted by M. Pasteur regarding the real causes of the chemical changes operative in alcoholic fermentation, I do not believe that the moment is appropriate to discuss them" (1859, p. 692). Traube echoes a similar view: "Schwann's hypothesis (later adopted by Pasteur), according to which fermentations are to be regarded as the expressions of the vital forces of lower organisms is unsatisfactory" (1878, p. 1984). Berthelot and Traube found Schwann's and Pasteur's views vitalistic because they did not advance a purely chemical model, but inserted living organisms as entities performing functions not capable of being performed by nonliving entities. These critics insisted on restricting their focus to purely chemical agents.

Those who rejected the claim that there was something unique about living cells that enabled them to accomplish fermentation generally embraced the view that there was a chemical agent within yeast that induced fermentation. The positive results with yeast could thus be explained. Berthelot, for example, compared fermentation to processes that were recognized as purely chemical, such as the inversion of cane sugar (that is, the hydrolysis of sucrose, yielding dextrose and fructose). While inversion of cane sugar was carried out by yeast, Berthelot traced it to a substance he could isolate from the yeast cell by precipitating a cell-free extract in alcohol and then redissolving it in water. He proposed that living organisms contained both soluble and insoluble ferments, the latter of which remained closely bound to the cell structure and could not be isolated from it. The chemical reactions, he argued, were due to these ferments, although, with Liebig, he assumed these were produced by the living organism.

The idea that fermentation was due to a localized substance within yeast was also maintained by Traube:

> Nothing points against the assumption that the protoplasm of the plant cell either is itself or contains a chemical ferment which promotes the alcoholic fermentation of sugar; and further, that the activity of this ferment only appears to be tied to the cell in so far as no means have been found up to the present of isolating the ferment from the cell without destroying it. In the presence of air, the ferment oxidizes sugar by transferring free oxygen onto it; in the absence of air, it decomposes the sugar by transferring oxygen from one atomic group of the sugar molecule to another, yielding alcohol as a product of reduction and carbon dioxide as a product of oxidation. (1874, p. 886)

Thus, both Berthelot and Traube continued to advance proposals for direct localization. Supposedly biological processes were to be localized as the effects of chemical agents within the cell.

Pasteur rejected the distinction between soluble and insoluble ferments. He maintained that anything that could be isolated from the cell and produce chemical reactions should be excluded from the class of ferments. According to him the association with living cells constituted the mark of a ferment. Others were, of course, more ecumenical in their use of the term.

4. CONCLUSION: SETTLING FOR DESCRIPTIONS

There is a common pattern to the four cases we have discussed in this chapter. Mechanists would offer models that appeared overly simplistic to those fully enmeshed in the complexities of the phenomena. Recognizing that the available mechanistic models could not deal with the phenomena that their experiments revealed, the opponents maintained the uniqueness of the phenomena. Often these opponents of mechanism were correct in pointing out the inadequacies of the then-current mechanisms, although wrong in their contention that no mechanism would ever suffice. Many of the appeals to vital forces that we have seen resulted from an inability to answer questions of interest at the level of organization at which research was then focused. Sometimes the appropriate level of explanation would involve a lower level, at which researchers then lacked the tools for carrying out empirical inquiry. Thus, Pasteur saw that explanations of fermentation in terms of underlying mechanisms were little more than crude speculation. In terms of the chemistry of the day, it could not be done. Similarly, Flourens saw that explanations of intelligence in terms of brain structure were unsupported by what was independently known of brain structure and function at the time. A systematic and principled account of the mechanisms was lacking. And it still is.

Almost as often, the appropriate level of explanation is higher than the one on which mechanists did focus research. We will discuss this in more detail in Chapter 7, but it can be briefly illustrated by Bernard's investigations into digestion and nutrition. Bernard began his research accepting the basic chemical models of digestive and nutritional processes common in the mid-nineteenth century. He assumed that there were highly localized chemical agents, catalysts, responsible for the processes in question. For example, Bernard's (1855) research leading to the glycogenesis of the liver was initiated under an assumption stemming from Liebig that all chemical reactions in the animal were catabolic, breaking down the complex foodstuff produced by plants into simpler products, with a release of energy. Thus, he sought the locus at which sugar was broken down within the body. Bernard planned to attack this problem and localize the reaction through a series of experiments in which he injected grape juice into animal blood systems and traced levels of concentrations by dissecting the animals at various times afterward. This led to the anomalous finding of heightened concentrations of sugar in the portal vein, upstream from the liver. Bernard then experimented with a dog kept on a protein diet and found the same result. Since it could not have originated in the animal's diet, Bernard asked from where the sugar came. He concluded that sugar was actually being formed in the liver from simpler constituents and was flowing back into the portal vein during dissection. This discovery undermined the simple, highly localized model that portrayed animal nutrition as involving simple assimilation and direct combustion. Nutrition had to be a much more complex process, involving both breaking down and rebuilding organic compounds; and it was clear that to develop a full account of nutrition one would have to examine the indirect path of chemical changes followed in the animal. This meant that nutrition could not be explained in terms of simple, basic, chemical reactions. It required, instead, an understanding of a complex sequence of reactions.

Arguments for vitalism often stem from researchers seeking an explanation at an inappropriate level, which may be either too high or too low; likewise, arguments for localization and decomposition can suffer from the same problem. When researchers are operating at too high a level, they cluster together too many activities into one entity, so that it becomes mysterious how that entity is able to carry out the functions assigned to it. When researchers are at too low a level they may miss some of the complexity in the organization and behavior of the system that is critical to understanding how the system actually accomplishes its activity. The chemists seeking to explain metabolic processes in the nineteenth century were both at too high and too low a level: They clustered different physiological functions together, creating tasks that were too complex for any single chemical agent; thus, they viewed fermentation as a simple catalyzed reaction when it was actually complex. On the other hand, they tried to handle functions like nutrition at too low a level, thus missing many of the interactive activities needed to explain it.

The cases described in this chapter were not chosen because they show ways the mechanistic program expands or grows, but rather because they point to a common way in which researchers have opted out of that program. Two features unify these cases and point to a common rationale for rejecting mechanism. The first is that opponents of mechanistic models generally were responding, in part at least, to real limitations on mechanistic theories. They recognized limitations on the resources offered by the mechanisms, and saw that they could not offer explanations of the phenomena. The second is that the systems are complex and require sophisticated mechanisms, which at the time seemed unavailable.

Although the case for abandoning mechanism is generally made *after* an attempt to pursue a localizationist program, the logical point of departure is *before* one enters into the direct localization program. The critical point is whether researchers elect to pursue a strategy of decomposing phenomena into component processes and localizing them in appropriate subsystems. The opponent of mechanism denies the utility of localization and decomposition and seeks instead a better understanding of the phenomena. Once again, we can portray this as an additional choice-point, as in Figure 5.1.

The decision to abandon mechanism is not without its own cost. For example, the inherent weaknesses in Flourens's own positive program offset the strengths in his critique of Gall. In a memoir reviewing Gall's and Spurzheim's anatomical works, the critics, who include Cuvier, make the following remark:

> We cannot expect a physiological explanation of the action of the brain in animal life. . . .
>
> The functions of the brain . . . consist in receiving, by means of the nerves, and in transmitting immediately to the mind, the impressions of the senses; in preserving the traces of these impressions . . . ; [and] lastly, in transmitting to the muscles, always by means of the nerves, the desires of the will.
>
> Now these three functions suppose the mutual, but always incomprehensible influence of the divisible matter and the indivisible mind (*moi*); a hiatus in the system of our ideas never to be supplied, an eternal stumbling-block of all our philosophies. (Tenon et al. 1809, p. 38)

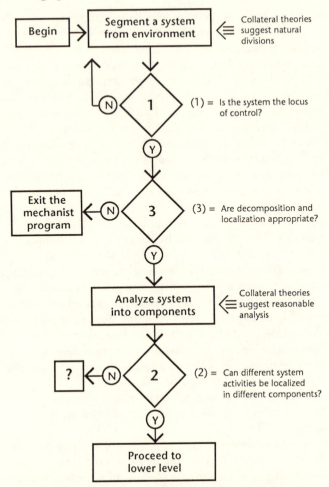

Figure 5.1. A Third Choice-Point. One response to attempts at mechanistic explanation involves a categorical denial of decomposition and localization, emphasizing a systematic description of the phenomena. This constitutes a rejection of mechanism.

A positive program for doing more than describing the functions is simply absent. The mechanistic program, however crude it might have been in the hands of Gall and Spurzheim, at least left us with a direction for research. Their opponents left us with only a "hiatus" which was "never to be supplied." Lacking a systematic program of research, the antimechanistic program is virtually guaranteed to fail in gaining adherents.

PART III

Elaborating Mechanisms

In order to really understand something such as speech, which is peculiar to man, it will be necessary to find ways of recording from single neurons from outside the skull.

—D. Hubel 1979

Since the human mind resides in the brain, we cannot be satisfied with our explanations of human thinking until we can specify the neural substrates for the elementary processes of the human symbol system. Of these connections we know next to nothing. . . . We are in a position similar to that of 19th-Century chemistry, which had developed an extensive theory of chemical combination long before that theory could be linked to the physics of atoms.

—H. Simon 1980

INTRODUCTION

In Part 2 we explored the preliminaries to mechanistic explanation. We turn now to the process of developing and elaborating such explanations. This requires showing how an activity that is performed by a whole system is accomplished by having different components perform different functions that contribute to the task. This is where the heuristics of decomposition and localization most properly come into play. Decomposition involves establishing a division of labor according to which different tasks involved in the same overall process are identified. Localization entails a systematic and independent examination of the processes operating at the lower level and a demonstration that these processes perform the functions specified in the decomposition.

We saw in Chapter 5 that direct localization was often charged by opponents with being vacuous or spurious. The natural consequence of this charge was a rejection of mechanism and a shift to vitalistic or dualistic views. Yet those who sought at all costs to avoid vitalistic or dualistic results could not simply neglect the power and force of the objections. The direct localizations we considered in Chapter 4 *do* posit control at the lower level for the very phenomena they are meant to explain. It was for this reason, together with the fact that the functions being studied were not obviously mechanistic, that the explanations were regarded as spurious. Put another way, direct localization was viewed as *insufficiently constrained localization*. In Gall's program, for example, higher-level cognitive capacities were localized in physical organs, but the only reason he offered for attributing specific capacities to specific organs was the need to explain the capacities. This is part of what we meant to imply by calling the evidence for Gall's view *merely* correlational. Flourens saw that this was not yet an explanation, and certainly not an explanation in terms of lower-level mechanisms. Similarly, Pasteur saw that the attempt to explain fermentation in terms of catalytic processes depended on showing that fermentation could be carried out in the absence of living organisms, and his classic experiments were designed to show that it could not. If fermentation could not be performed without living organisms, then the postulated catalysts could not be simple chemical agents, and the appeal to chemical mechanisms for fermentation could offer no serious explanation. It is the lack of *independent* constraints on what functions are localizable that makes the solutions *insufficiently* constrained.

The point is a general one. We are told that the capacity for language comprehension is localized in regions posterior to the Sylvian fissure; that the cerebellum is the locus for motor control; that the left hemisphere is specialized for "analytic" tasks (while the right is more concerned with pattern and synthesis); and that there are genes for altruism, parental care, manic depression, schizophrenia, and Alzheimer's dementia. These are all claims to localization, but they are by no means all on an equal footing with respect to either evidential or theoretical credentials. Some are well motivated. Some are not. One factor that is important for our purposes in Part III is *the character of evidence* that serves to underwrite such claims. A well-developed localizationist model depends on detailing the mechanisms, robustly characterized. It is important that we can detail the mechanisms, and it is important that the evidence renders them robust. In some cases, such mechanisms can be isolated and characterized; in others, they cannot.

Let us consider two relatively recent examples. There have been highly publicized accounts claiming to have isolated a gene for manic depression (within a small population) on a single chromosome (see Egeland et al. 1987). It has long been accepted that manic depression has a significant hereditary component. More recently, researchers have claimed to isolate genetic markers on the tip of the short arm of chromosome eleven that are common among those members of an Amish community in southeastern Pennsylvania who suffer from manic depression. As the researchers describe it, there is a linkage between these markers and "a locus conferring a strong disposition to bipolar affective disorders" (ibid., p. 784). The researchers are, of course, under no illusion that there is a single gene controlling manic depression, and they explicitly allow that other genes are involved as well as that there is heterogeneity in manic depression's genetic basis; nonetheless, they claim that there is a correlation between a specific chromosomal pattern and the presence of manic depression within that Amish community and that, within that population, this chromosomal pattern is indicative of a gene resulting in manic depression. It is important to recognize how tightly circumscribed this claim is. There is no claim to know the mechanism underlying the etiology of the disease. The data is limited to a specific population; moreover, there is obvious evidence that this correlation is limited in scope.[1] This is a clear claim to localization, though the underlying mechanisms are admittedly complex. What we wish to underscore is that the *evidence* for the claim is correlational only. Independent physical constraints are simply lacking. We do not know how the correlated chromosomes could constitute causal agents in the etiology of the disease; in fact, we do not even know *whether* the genes identified are causal agents or just simple markers.

It is interesting to compare this claim with another recent, and analogous, claim to have localized the defect responsible for Alzheimer's disease (see Barnes 1987). This latter claim suffers few limitations paralleling those present in the research into manic depression. On the one hand there is evidence from one group of researchers (Goldgaber et al. 1987) that indicates the defect in at least one form of Alzheimer's disease—the inherited variant—is located on chromosome twenty-one. This evidence is drawn from information on four distinct family groups. Other research locates a gene coding for an amyloid protein in the proper region on chromosome twenty-one (Tanzi et al. 1987), and yet another group reports an accumulation of amyloid protein filaments in the brains of aged mammals and an unusually high deposition of this protein in Alzheimer's patients (Selkoe et al. 1987). Finally, there is clear evidence that in Alzheimer's patients there is a gene duplication not found in control subjects that would, presumably, result in an exaggerated dosage of the amyloid protein (Hyman et al. 1984). All of this is hardly conclusive and is insufficient to determine the precise etiology of Alzheimer's disease. We do, though, have solid reasons to think that amyloid proteins are involved, even if we do not yet see exactly how.

What is important for our purposes is that this case not only provides evidence of discrete localization, but also connects this localization with evidence suggestive of a mechanism—or, more precisely, with evidence suggesting that there is a determinate genetic mechanism. An impressive array of evidence converges on the same result. What is notably lacking in the study of the Amish community is any biochemical clue as to why the genes are implicated in the disease, or how they might contribute to it. The Alzheimer's case, by contrast, offers robust evidence for the localization, and some of the evidence is at least indicative of an underlying mechanism.[2] We will see that this sort of coupling, drawing on independent theories and evidence at more than one level, is important to localizationist research.

The cases to be described in Chapters 6, 7, and 8 all share one important feature not present in the cases of Part II: lower-level theories, and empirical results accessed at those lower levels, impose significant constraints on the development of the relevant theories. That is, in the terms we used in Chapter 1, the lower-level theories provide constraints that limit the search space. It turns out that these constraints are powerful. They provide information about the class of allowable mechanisms and so limit the space of mechanisms needing to be searched (much as the assumption that relations between variables were linear aided BACON in inducing Boyle's Law). The strength of these constraints—the extent to which they effectively limit the search space—varies. In Chapter 6 the constraints are relatively weak and fundamentally empirical. In Chapters 7 and 8 they are substantially stronger and have a theoretical component as well as an empirical one. We will also see that many of the constraints are modified or abandoned altogether. As more constraints are imposed, and the task is one of simultaneously satisfying independent constraints, then it sometimes becomes necessary to compromise some in order to satisfy others.

In Chapter 6 we examine work in the neurosciences. One central case is the classic work on aphasias by Karl Wernicke toward the end of the nineteenth century, and the second is a more recent psychobiological investigation of spatial memory carried out by John O'Keefe and Lynne Nadel (1978). On the basis of systematic deficits observed in patients, Wernicke concluded that there were localized centers for a variety of associations. One important feature of Wernicke's work, in contrast to that of, say, Broca, is that it was explicitly based on neuroanatomical evidence concerning the projections of cerebral tracts. O'Keefe and Nadel construct a model more severely constrained by the empirical data. After examining these cases, we will discuss the role of localization and decomposition in the interpretation of the experimental results and show that the localizationist program in its stronger forms requires a convergence of cognitive models and neural mechanisms.

Though O'Keefe and Nadel begin with a direct localization of objective cognitive and spatial maps in the hippocampus, they take up the task of explaining how the hippocampus performs this function. This requires a shift to a lower level. Accordingly, they focus on the components of the hippocampus and not on the whole organ. This introduces lower-level constraints. They appeal to empirical data detailing different patterns of neuronal activity in different parts of the hippocampus, including both single-cell recordings and EEG patterns. They then proceed on this basis to develop a functional analysis showing how the activities occurring in different components of the hippocampus at different stages in the performance of spatial-problem-solving tasks could accomplish the overall task of spatial mapping. This is the sort of model that we refer to as a complex localization. The data concerns the behavior of individual cells and patterns in their activity. The correlation of the activity of individual cells with differences in spatial-problem solving provides a more substantial empirical basis for localizing spatial representation in the hippocampus. The decompositional analysis of the hippocampus, and demonstration of how its components could carry out the necessary processing to be the seat of the capacity to develop and use objective mental maps of spatial layouts, serves two functions. First, it provides independent, and hopefully convergent, empirical evidence vindicating the initial direct-localization claim; and second, it offers a mechanistic account of how the task is performed.

Chapters 7 and 8 examine cases imposing still stronger lower-level constraints, including both empirical and theoretical components. Chapter 7 focuses on research devoted to determining the fermentation mechanism in living cells. As we observed in Chapter 5 fermentation was a process that seemed to elude mechanistic explanation in the nineteenth century. The major breakthrough came with Buchner's demonstration in 1897 that fermentation could be achieved in a cell-free extract. Despite his initial direct localization, researchers quickly realized that one enzyme could not accomplish the complex reaction of fermentation. Rather than developing the mechanistic explanation by shifting to a lower level and finding subcomponents that performed the function through their interaction, research remained at the same level and sought to identify a variety of components that would each play a different role in carrying out the overall task of fermentation. It was not possible to identify and distinguish enzymes physically, so researchers had to proceed functionally instead. They investigated various ways of inhibiting part of the functional pathway, and they stimulated the operation of the pathway by supplying potential intermediaries in an attempt to determine the stages in the fermentation process.

The first complex-localizationist models proposed for fermentation assumed that the process consisted principally of a linear series of reactions through which sugar was rendered into alcohol. In the language of Chapter 2, researchers assumed that the fermentation system was a component system and thus nearly decomposable, that each step in the reaction was brought about by a specialized enzyme and did not depend upon other steps. Eventually researchers realized that the various reactions in the pathway were closely linked through a variety of connecting mechanisms; that is, they came to realize that fermentation relied on an integrated system. Thus, Chapter 7 first examines the development of mechanistic explanations that begin with direct localization, then looks at the advancement of a more complex mechanism that treated the system as a component system, and finally introduces an account of an integrated system.

In Chapter 8 we turn our attention to biochemical genetics. This field brought together a tradition of research on biochemical pathways with research on genetic control, focusing on the mechanisms by which genes expressed themselves in development. Mendelian genetics had developed a complex account of the genetic mechanisms that mediated heredity. This was a localizationist scheme. However, the model could not explain how genes expressed themselves in traits. This was recognized to involve a set of chemical reactions building up the ultimate products. The research program of biochemical genetics was directed at determining the pathways responsible for expressing the genes already identified through the Mendelian program. Ephrussi, Beadle, and Tatum investigated the biochemical processes involved in developing normal nutritional processes in *Drosophila* and *Neurospora*. The *Drosophila* work indicated the complexity of genetic control in development. It was still research based on the Mendelian paradigm, attempting to discern genetic control from differences in the phenotypic expression. The *Neurospora* work provided information concerning the level of control. It began by inducing mutations and showing the phenotypic effects. It revealed, more specifically, that mutants lacked one or another component of the normal biochemical pathway of nutrition. Beadle and Tatum saw this as a new account of the complex mechanisms of genetic expression, on account which linked genes not with Mendelian traits, but with enzymes in biochemical pathways.

The result of Beadle and Tatum's analysis identifying genes with enzymes was that genes were no longer viewed as coding for Mendelian traits directly. Thus the localizationist story could not explain the expression of Mendelian traits, as it was originally intended. In this case, though, the consequence was not a rejection of localization and decomposition, but a *reconceptualization* of the phenomena to be explained. The research at the lower level forced a reconstitution of what researchers thought existed at the higher level; what seemed to be an appropriate account of the mechanism when all one had were the functional tools of Mendelian genetics was revised as a result of research directed at the lower-level analysis. We will see that specific lower-level constraints played a critical role in this transition.

CHAPTER 6

Complex Localization

1. Introduction: Constraints on Localization

Mechanism was often seen as the only viable alternative to vitalism, and it was undertaken despite the fact that the simple mechanistic approaches were not wholly satisfactory as explanations, even on their own terms. At the same time vitalism can be seen as a consequence of mechanism, drawing sustenance from a clear vision of mechanism's limitations. Thus the mechanism of Gall, or of the preformationists, was mechanistic, but hardly provided a palliative for the explanatory needs it was designed to fulfill. As Flourens saw very clearly, simply relocating the faculties at a lower level was not an adequate explanation in the absence of an account of how the organs could do what they needed to. Simple, direct localization is by itself no solution, though it can serve as a preliminary stage bridging toward a more adequate account. When additional constraints are imposed, whether empirical or theoretical, they can serve simultaneously to vindicate the initial localization and to develop it into a full-blooded mechanistic explanation. In a somewhat less paradoxical form, additional constraints can result in a model that is more defensible, displacing and developing the initial localizationist picture. What we call complex localization is *localization multiply constrained*; that is, it proposes a set of components that contribute differentially to system function, and it incorporates independent constraints on allowable mechanisms from lower levels.

In addition to depending on multiple constraints, complex localization requires that we take decomposition seriously. Simple localization differentiates tasks performed by a system, localizing each in a structural or functional component. Complex localization requires a decomposition of systemic tasks into subtasks, localizing each of these in a distinct component. Showing how systemic functions are, or at least could be, a consequence of these subtasks is an important element in a fully mechanistic explanation. Confirming that the components realize those functions is also critical. Both are necessary for a sound mechanistic explanation.

As we shall see, the routes to complex localization are varied. Frequently a program of research begins with direct localization, which then develops into a more complex localization in which functional decomposition of tasks assumes a more central role. This is common in psychology, where the legacy of Gall continues to influence theory development. Research often commences by dividing psychological activities into broad performance categories; for example, perception, memory, language, reasoning, and emotion. Noam Chomsky has provided one of the clearest expressions of this approach in his own "organology," strikingly reminiscent in tone of phrenology (cf. above, Chapter 4, section 2):

> We may usefully think of the language faculty, the number faculty, and others, as 'mental organs,' analogous to the heart or the visual system or the system of motor coordination and planning. . . . There seems little reason to insist that the brain is unique in the biological world in that it is unstructured and undifferentiated, developing on the basis of uniform principles of growth or learning— say those of some learning theory, or of some yet-to-be conceived general multipurpose learning strategy—that are common to all domains. (1980, p. 39)

Jerry Fodor has articulated the modularity, or organology, and generalized it beyond the domain of language, claiming that "modular cognitive systems are domain specific, innately specified, hardwired, autonomous, and not assembled" (1983, p. 37). These modular systems are concerned with a single domain and do not depend on the operation of other capacities in the process of executing their own specific functions. It is in roughly this sense that, for example, it has been held that syntactic processing does not depend on semantic interpretation, or that early visual processing is autonomous. Gall's phrenology also fits the profile as a theory postulating a modular structure to the mind. Destructiveness and combativeness are distinct faculties, after all.

Modularity seems inevitable, especially when dealing with a phenomenon as complex as mental life. If we do not divide up mental activities, or cannot discern their variety, we cannot even begin to see how they are realized. However, even when modularity is correct as a first approximation, it is only a beginning. It is imperative to know how the mechanisms work, and it is certainly not unreasonable to turn to the neurosciences to develop mechanistic models of psychological activities. It is in fact hard to see where else one might turn to gain a robust and realistic account of the mind. In this chapter we look at two examples from the neurosciences, one from the nineteenth and one from the twentieth century. Both are unrelentingly mechanistic.

It is important to recognize the eclectic character of neuroscience. There are variations in general strategy, in substantive methodology, as well as in domain; and all influence the character of the resulting models of cognitive functioning. Researchers vary, in particular, on the relative emphasis they place on cognitive models of performance as opposed to neurophysiological analyses. Some researchers are inclined to take cogni-

tive categories as relatively fixed and view the problem to be one of identifying neural structures responsible for the capacities those categories
describe. For example, Norman Geschwind (1964, 1965, 1980) has elaborated a model for the neural substrate of language which emphasizes cognitive categories. He suggests that there are two primary structures implicated in speech: one concerned, roughly, with speech comprehension,
and one with the control of speech output. Other researchers by contrast,
are inclined to emphasize models based on information pertaining to neurophysiological processes. Thus Eric Kandel (Hawkins and Kandel 1984)
and his collaborators have developed a model of learning in *Aplysia* which
is motivated largely by detailed work on neurophysiological organization
and, secondarily, by views about the mechanisms of associationistic learning. Likewise, Daniel Alkon (1989a, b; Farley and Alkon 1985) and his
associates have a model of learning in *Hermissenda* which traces variations
in learning to neural systems and associated cellular changes.[1] These differences are of emphasis and of degree, but they are differences that
matter.

Even if the models depend on cognitive categories that are tentative
and modifiable, they still significantly guide the search for neural mechanisms. The goal is, after all, understanding how the tasks *already identified* are performed. This does not deny that neurophysiological results
might lead to the elaboration or modification of cognitive categories; they
might even lead to wholesale revision or replacement, though we think
this is certainly infrequent (cf. McCauley 1986; Wimsatt 1976). The point
is, rather, that the models motivated by neuroscience cannot ignore cognitive psychology if the goal is understanding how cognitive tasks are performed; neither, despite philosophical imprimatur, can psychology ignore
the neurosciences.

The solution is to seek robust, multiply constrained models. Psychological categories—or, better, *psychological organization*—provide a top-
down constraint. Different conceptions of the proper decomposition of
psychological functions impose different constraints on mechanistic models. Wernicke and Broca, for example, differed radically on which psychological organization was important. Broca embraced a faculty psychology,
while Wernicke depended on an associationistic psychology. As we will
see in the next section, an associationistic psychology leads to different
mechanisms, or to different ways of conceiving of mechanisms, at the
lower level. *Neural mechanisms* provide a bottom-up constraint, limiting
the range of possible mechanisms that can explain psychological functions.
Minimally this requires some consistency with what is known of neuroanatomy and neural pathways; more ambitiously it requires knowledge
of neural mechanisms that have an appropriate internal organization.

Complex localization, at least in the cases we will describe in sections 2 and 3, results from an attempt to satisfy, simultaneously, constraints imposed at more than one level. This is what we mean by localization multiply constrained.

2. TOP-DOWN CONSTRAINTS

The differences, including the gains from multiple constraints, can be made clearer by returning to direct localization, and, in particular, to the natural heirs of Gall and direct localization. One of the more prominent French physicians to embrace the commitment to localized centers within the brain was Jean Baptiste Bouillaud (1796–1881), a student of Gall, founding member of the Société Phrénologique, and professor of clinical medicine at La Charité hospital in Paris. In 1825 Bouillaud described a patient who had lost the power of speech after a lesion in the anterior lobe, arguing that this was a direct confirmation of Gall's view that this lobe was the locus for articulate language. Bouillaud's method, very much like Gall's, relied on correlations; but, unlike Gall's, compared pathological symptoms with neurological lesions: Bouillaud regarded the examination of the dysfunctional brain as essential. His views met with considerable opposition, and after many public defenses Bouillaud was led to offer five-hundred francs to any person who could produce a case with no disturbance of speech in the presence of a lesion in the anterior lobes of the brain. No one collected on the bet.

This was the climate into which Paul Pierre Broca (1824–1880), as a defender of organology, entered the debate over localization. Although he rejected the cranioscopic method of phrenology, he did not reject the traditional emphasis on the size of the brain as indicative of intelligence, nor did he reject the reliance on cranial capacity in comparing races. He preferred, though, to measure brain capacity more directly, and he emphasized the pathological and clinical observation adopted by Bouillaud. Like Gall and Spurzheim, Broca was committed to direct localization. In sessions of the Société d' Anthropologie in 1861, and shortly thereafter at the Société Anatomique de Paris, Broca praised Gall and the principle of localization of function, describing it as "the point of departure for all the discoveries of our century on the physiology of the brain."

By August of 1861, Broca had described in considerable detail the anatomical changes accompanying a disorder of speech which he termed aphemia: a disturbance of "articulated speech" in the absence of any disturbance of the "general faculty of speech." The subject of the study came to be known as "Tan," following his habitual answer to questions. A congenital epileptic, Tan lost the ability to speak by the time he was thirty, but was otherwise healthy and intelligent. Over the ensuing years, however, Tan's case degenerated, with weakness and eventual paralysis of the right side of his body. Eventually, he died from gangrene. Through interviews with the hospital staff and other patients, Broca concluded that Tan's initial aphasia amounted only to a "loss of articulate language," with no comprehension deficit, no loss of hearing, and no loss of intelligence. Broca summarized the case as follows:

> From the anamnesis and from the state of the patient it was clear that he had a cerebral lesion that was progressive, had at the start and for the first ten years remained limited to a fairly well-circumscribed region, and during this first period had attacked neither the organs of motility nor of sensitivity; that after ten years the lesion had spread to one or more organs of motion, still leaving unaffected the organs of sensitivity; and that still more recently sensitivity had become dulled as well as vision, particularly vision of the left eye. Complete paralysis affected the two right limbs; moreover, the sensitivity of these two limbs was slightly less than normal. Therefore, the principal cerebral lesion should lie in the left hemisphere. (1861b, pp. 226–27)

An autopsy revealed widespread damage to the left frontal lobe. The tumor consumed the superior part of the frontal lobe, as well as much of the temporal lobe and other structures. As Broca described it, "If one puts back in imagination all that has been lost, one finds that at least three quarters of the cavity has been hollowed out at the expense of the frontal lobe" (ibid., p. 227). Assuming that the tumor spread uniformly, Broca concluded that its initial focus had to be in the frontal lobe, in particular in the third convolution of the frontal lobe (see Figure 6.1). The correlation of the symptoms and the projected course of the tumor led Broca to conclude that Tan's initial speech deficit was due to localized damage to the third frontal convolution and that, since other faculties were not initially disturbed, this must be a region specialized for language.

Embracing the "principle of localization" inherited from Gall, Broca boldly concluded from Tan's case that "the totality of the convolutions does not constitute a single organ, but many organs or many groups of organs, and there are in the cerebrum large discrete regions corresponding to large discrete mental functions" (ibid., p. 227). This is, of course, nothing less than Gall's organology. This famous case has earned Broca the reputation as the first to isolate a case of pure aphasia and to gain general acceptance in localizing a specific function in the brain (cf. Young 1970, ch. 4). His central contribution was probably a clear description of the phenomena associated with the lesions.[2]

It would not be much of an understatement to say that there was no novel clinical syndrome exhibited in the case described by Broca, and that the anatomical basis for his cortical localizations was meager indeed. Eventually, as a result of superior precision and technology, both meth-

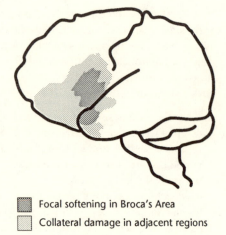

Figure 6.1. A Representation of the Damage to the Left Hemisphere in Tan, According to Broca (1861). Broca concluded, on the basis of the progress of Tan's aphasia and a post-mortem examination, that the damage in the focal area of the brain was responsible for Tan's condition and that that region was the "center for articulate language."

▨ Focal softening in Broca's Area
▢ Collateral damage in adjacent regions

ods and answers improved. By examining Wernicke's views, and then more recent work on spatial memory, we should be able to see one path toward more complex localizationist models. This will occupy us in what follows.

Wernicke on Language

Karl Wernicke's (1848–1905) work on sensory and motor aphasias is one of the clearest cases of research exhibiting the heuristics of decomposition and localization. Wernicke assumed that there was a series of discrete loci mediating the comprehension and production of speech; loci which, in concert, control behavior. There is, of course, interaction between these centers in the normal case: comprehension and expression are coordinated. The critical question from Wernicke's perspective, however, was whether it was *possible* for there to be normal comprehension in the absence of normal expression, or vice versa. He thought that the subsystems were independent despite their interaction, meaning that each could carry out its usual operations independently of the other; comprehension *could be* gained in the absence of expression, and expression effected in the absence of comprehension. Wernicke went on to assume that the subsystems also operated independently of one another in individuals not suffering from aphasia, that comprehension was actually independent of expression. The specific character of these *association centers*, and the functions they subserve, were interpolated on the basis of the specific clinical disabilities that resulted from injury to these centers. These behavioral deficits were viewed as the empirical cues to assessing psychological organization.

As we have pointed out, it is difficult to identify mechanisms when systems are operating smoothly, because components do not reveal themselves. The way the system breaks down often provides the best information available about the functional properties of the components. Thus, it is not merely that behavioral deficits are clues to *abnormal* functioning—they are clues to *normal* functioning as well. Functional independence, or decomposability, shows up if some capacities can be lost while others are unaffected. For example, loss of comprehension without a corresponding loss of expression, or vice versa, supports the claim to independence. On the other hand, there would be a failure of functional independence if there were *always* comprehension deficits pursuant to damage to the subsystem responsible for expression.[3]

These centers of comprehension and expression, however, were not the faculties posited by Gall and Broca. Wernicke assumed an associationistic model of learning and memory. Concepts are not independent entities, according to him, but the product of linking together more basic representations; the psychological unit becomes a set of associated ideas rather than an independent capacity. In Wernicke's own formulation, the possession of a "concept of a concrete object" is identified as a "constellation of memory images"; that is, it is a cross-modal association.[4] A concept of an observable object is thus identified with an association either between sensory representations, or between sensory and motor representations. Sensory impressions, or sensory representations, are modality-specific "ideas" or "representations" (*vorstellungen*), and they correspond to classical "proper sensibles." The concept of a rose, Wernicke says,

> Is composed of a 'tactile memory image'—'an image of touch'—in the central projection field of the somesthetic cortex. It is also composed of a visual memory image located in the visual projection field of the cortex. The continuous repetition of similar sensory impressions results in such a firm association between those different memory images that the mere stimulation of one sensory avenue by means of the object is adequate to call up the concept of the object. (1906, p. 237)

Given an associationistic slant, what are localized must be the centers of association rather than the cognitive faculties. A concept will not be localizable, since it will be no more than a constellation of associations; a faculty will not be localizable, since there are no faculties. The centers of association alone will possess discrete loci, and disruptions of these loci will in turn disrupt some aspects of behavior. In the case of language we must seek the associations built up in the process of learning to comprehend and produce speech; in speech perception, the simple ideas are acoustic representations; and in speech production, the ideas are the motor representations of speech (*Sprachbewegungsvorstellungen*)—that is, they are the representations of the movements necessary for speech. Comprehension or expression will depend on the establishment of associations between simple ideas. These associations vary in complexity, from the primary reflex movements controlled by subcortical representations, to secondary motor movements initiated by cortically based representations, to voluntary control of behavior at the highest level (cf. Wernicke 1874, p. 95).[5]

Learning is, at least to a first approximation, understood as establishing an associationistic link between sensory impressions and motor representations. Accordingly, memory is a functional association between perceptual elements resulting from lowered resistance—or facilitation—following repeated, contiguous presentations. These include the sound/concept associations that are constitutive of sensory representations, the concept/word associations constitutive of motor representations, and the sound/word associations constitutive of inner speech.

The associationistic psychology embraced by Wernicke provided a psychological mechanism in terms of which aphasic syndromes could be interpreted. Any "higher" cognitive performance would be the result of an association of simple ideas. The conception of the psychological mechanism thus provided an account of what needed to be localized. The primitive ideas themselves, and especially the pathways that accomplished the association between them, were to be localized. To demonstrate that these were localized agents performing independent functions, Wernicke appealed to the different kinds of linguistic deficits (the aphasias) found in language comprehension and production. According to this account, specific aphasias were explained as breakdowns in an associationistic system; the diversity of the aphasias was a natural consequence of the complexity of this system.

Wernicke's approach allowed a systematic treatment of the various forms of aphasia within a single model, and that treatment provided the primary motivation for his model of language comprehension and production. It had been known for some time that localized destruction could impair comprehension of speech, while leaving both production and more general intellectual functions unimpaired. Wernicke noted a parallel phenomenon in which localized destruction could impair production of speech, while leaving both comprehension and general intellectual functions unimpaired. He saw that disruption of the first frontal gyrus would disrupt production; of the first temporal gyrus, comprehension (cf. 1874, p. 103). The former affliction has come to be known as *Broca's Aphasia*, and the latter as *Wernicke's Aphasia*.

Of course, in the vast majority of cases, what we see is an impairment in both categories as well as a variety of other cognitive deficiencies. Wernicke knew this well.[6] As he saw, however, the existence of complex deficiencies is less important than the existence of the "pure aphasias"; that is, aphasias that affect strictly limited associations while leaving others intact. This is evidence that the representations can be dissociated, and therefore is evidence for decomposability within the system responsible for language. What most notably distinguishes Wernicke's program from Broca's, is this interpretation of the two primary forms of pure aphasia as sensory and motor deficiencies. As Wernicke understood sensory aphasia, there is a loss of the "acoustic memory representations," which provide the sound-concept associations. The meaning of heard speech therefore cannot be recovered. Articulate speech is intact; indeed it is copious and grammatical, though inappropriate to the context (cf. Wernicke 1874, p. 124). In motor aphasia there is a loss of the "motor memory representations" for speech, which provide the concept-word associations. Comprehension of speech is intact, but there is a loss of articulate speech (ibid., pp. 126, 130). In either case—for Wernicke, but not for Broca—the problem is not the loss of the concepts or the simple ideas, but an inability to establish the necessary associations.

As Wernicke saw it there are a variety of forms of pure aphasia, each with a characteristic syndrome and with very little overlap in symptomology. He wrote, even as early as 1874:

> It is self-evident that cases presenting two or three symptom-pictures in combination occur more frequently than the pure clinical forms caused by more or less isolated damage because the pathological process is generally very extensive. Undoubtedly, the typical pictures described, which adequately justify our formulation of a new clinical classification, actually do exist. . . . The mixed forms may readily be understood from the preceding discussion. (p. 114)

Clinical classification reflects the possible. The actual is more complex. Wernicke infers that there are centers responsible for these simple functions shown by the forms of aphasia, and these alone.

Wernicke's approach shows clearly how decomposition and localization depend upon one another. Information about the effects of localized disruptions shapes our conception of our basic psychological capacities. Clinical data is the motivation for distinguishing comprehension from expression, while circumventing the charge that the distinction between the two is but a false abstraction. On the other hand, views concerning psychological organization provide an account of what needs to be localized. Wernicke was limited to clinical data because he concentrated on human subjects in developing his model. There are other capacities that, within the last century, have become approachable with more sophisticated experimental techniques. It will be useful to look at a more recent case of localization based on an experimental rather than a clinical methodology. The general moral for our purposes is not substantially different.

Spatial Memory

In an interesting and impressive comprehensive study of hippocampal functioning, John O'Keefe and Lynn Nadel (1978) urge that the hippocampus is a structure specialized for spatial processing and memory. This is by no means the only proposal for understanding hippocampal functioning (for a general review, see Isaacson 1974), and O'Keefe and Nadel's proposal is far from unproblematic. In particular, David S. Olton and others have argued that the critical factor in hippocampal functioning is its involvement in working memory, rather than any involvement in spatial functioning (e.g., see Olton, Becker, and Handelmann 1979). It is not important to our project which, if either, of these proposals is correct, or which might finally be vindicated by further research. What is important is the methodological moral. O'Keefe and Nadel pursue a robustly characterized mechanism underlying spatial memory. They propose not only a localized capacity and an analysis of the relevant capacities, but a physical mechanism. As a result they propose a model that integrates information gleaned from more than one level and is subject to multiple, independent constraints.

O'Keefe and Nadel begin at a psychological level with the problem of explaining capacities for spatial orientation in humans and other animals. They distinguish two cognitive systems for guiding behavior; these systems provide alternative methods for solving spatial problems. What they call the *taxon system* provides an observer-relative, or egocentric, map of the environment. Activity guided by the taxon system is, accordingly, cued to external guideposts and bodily orientation. For example, an organism may elect to turn left at a given landmark in a maze when coming from a particular direction, just as we may be instructed to turn left at the second corner past the light. The *locale system* provides an observer-independent, or objective, map of the environment. Orientation is within an objective space. An animal may approach a position in the environment and, independent of the direction from which it approaches or any specific landmarks, still know its way around. For example, migratory birds can return to nesting grounds without depending on local cues, just as we may find our way to Chicago whether we begin in Minneapolis or Cincinnati.

An example used by O'Keefe and Nadel may help to introduce some operational rigor into the distinction. Tolman, Ritchie, and Kalish (1946) provided evidence for the deployment of objective cognitive maps by rats using a sunburst maze (see Figure 6.2). In the initial, training, phase (Fig. 6.2[a]), rats were taught to run a maze through a central chamber and along a side passage to a goal box with feeding stalls. After this phase the rats were released in a sunburst maze (Fig. 6.2[b]), which allows an array of directional choices. A *taxon* orientation would dictate that the rats se-

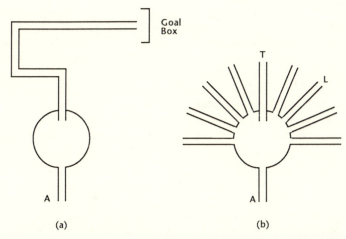

(a)   (b)

Figure 6.2. A Simplified Version of a Sunburst Maze Used by Tolman,
Ritchie, and Kalish (1946). The rats were trained initially on the simple maze
depicted in (a), taught to run through the central chamber from A, through
the corridor, and to the goal box at the end. Subsequently, rats were released in a sunburst maze such as that illustrated in (b). Tolman, Ritchie,
and Kalish reasoned that the path chosen in the second phase would reveal
whether the rats had learned a specific route to follow or had learned the
location of the goal. Rats uniformly took the direct route in the later trial.

lect the same passage (path T) they had been trained to use, as this would
maintain their relative orientation and preserve response invariance. A
locale orientation would require the selection of one of the paths providing a direct route toward the stimulus (ideally, path L), as this would reflect an awareness of location independent of immediate cues. In the experiment, the rats worked with a local orientation, leading Tolmon,
Ritchie, and Kalish to conclude that rats had spatial maps.

O'Keefe and Nadel differentiate the two systems by turning to behavioral deficits associated with localized damage. They contrast two cases:
"normal" animals (those which have undergone no surgical ablations) and
"hippocampal" animals (from which the hippocampal structures have been
bilaterally destroyed). O'Keefe and Nadel suggest that normal animals use
both systems, while hippocampals use only a taxon system. This is based
on the supposition that the locale system resides in the hippocampus,
while the taxon system is localized in the parietal cortex.

The hippocampus, then, is the source of the *cognitive maps* that represent a nonegocentric, local-driven space (O'Keefe and Nadel 1978, p.
382). To make good on this claim, O'Keefe and Nadel must functionally
dissociate the two, alternate, memory systems. In doing so they point to
cases in which hippocampals perform less well than normal animals, em-

phasizing differences in the patterns of errors (1978, p. 382). In arguing for the claim that the hippocampus provides a cognitive, observer-independent map, it is crucial for the project that there be substantial decomposability. O'Keefe and Nadel explain their strategy in suggesting that "the hippocampus acts as a cognitive mapping system, which generates place hypotheses and exploration. Loss of this system forces the animal to rely on the remaining extrahippocampal system" (ibid., p. 90). The lesion studies that form the first, top-down leg of their analysis of hippocampal function depend on the assumption that the taxon system can function properly in the absence of the locale system (and presumably vice versa) and on a disparity between the performance of normal and hippocampal animals—that is, between normal animals and those with bilateral removal of the hippocampus and related structures. Assessing the difference between these animals depends critically on there being two functionally different cognitive systems controlling spatial orientation and on the system's discrete localization.

It is clear that, relative to normal rats, hippocampals are at a systematic disadvantage in running mazes. Although they uniformly perform worse than intact animals, it is *not* the higher frequency of gross errors that is interesting. It is not at all remarkable that, after extensive damage to their brains, animals with bilateral ablations do not function up to par. What is of special interest is the pattern of errors they manifest. For example, O'Keefe and Nadel cite a number of studies using mazes requiring spatial reversals; that is, "complex" mazes that demand, for completion, that the animal turn directly away from the goal (cf. O'Keefe and Nadel 1978, pp. 288ff.). Performance on mazes of this sort can be compared with performance on "simple" mazes which require no turns at more than right angles to the goal. A normal animal will be under the guidance of a locale system which contains an objective map of the environment. It will then orient toward the unseen goal. A normal rat's reliance on a locale system should make solution to the complex maze comparatively more difficult. It should follow that the rate of errors on complex mazes should be *greater* than on the simple mazes. But suppose, on the other hand, that a hippocampal animal lacks such an orientational capacity. It must therefore, following O'Keefe and Nadel, rely on taxon systems which are insensitive to objective orientation. We should then find no greater *relative* frequency of errors on the complex mazes, however great the gross number of errors may be. That is, hippocampals should perform worse than normals on both simple and complex mazes, but should show no significant difference between the two. And indeed, this is exactly what we find.

In one study by D. P. Kimble, the normals made three times as many errors on the complex as on the simple maze, even though the opportunity for errors was greater on the latter. Hippocampals, by contrast, though significantly impaired on both mazes when compared to the controls, made twice as many errors on the simple mazes in comparison to the complex mazes. It would appear—or so O'Keefe and Nadel conclude—that hippocampal animals do not rely on the same hypotheses in problem solving as do the normal animals, and, lacking these hypotheses, they are less prone to error in cases where such hypotheses would be counterproductive; or, more carefully, the hippocampals' proportion of errors will not increase in circumstances that would make behavior based on the hypothesis prone to error.

The localization of the locale system in the hippocampus depends on the view that the functions of memory are decomposable and can be dissociated so that, in some cases, they function independently. O'Keefe and Nadel are explicit on the point:

> It appears that there are different types of memory, relating perhaps to different kinds of information, and that these are localized in many, possibly most, neural systems. . . . [There] is no such thing as the memory area. Rather, there are memory areas, each responsible for a different form of information storage. (1978, pp. 373–74)

Complexity once again is treated as multiplicity.

O'Keefe and Nadel thus assume there is decomposability at more than one level. First, there must be substantial independence of systems involved in memory from other, possibly related, cognitive functions. If damage to the locale system impaired general cognitive functioning, we would expect to see a more complex mosaic of symptoms than hippocampal damage alone would suggest. Second, there must be substantial independence of the various components of memory from one another. The taxon system is supposed to execute its characteristic functions even in the absence of structures that realize the locale system.

Because O'Keefe and Nadel claim that different neural mechanisms are involved in these two spatial systems, it is critical that the two systems are both functionally independent and localizable in different neural structures. This is straightforwardly inconsistent with equipotentiality of neural structures. Equipotentiality would allow alternative regions of the brain to assume functions they do not normally realize and so would undermine the specialized character of the posited mechanisms. For example, young children (from roughly 20 to 36 months) who suffer traumatic aphasia, effectively recover language skills. "In the very young," Eric Lenneberg tells us, "the primary process in recovery is acquisition, whereas the process of symptom-reduction is not in evidence" (1967, p. 150). If there were a potential for recovery of function in this same sense for memory as well as language, then any appeal to the neural apparatus specific to the hippocampus would be pointless, for there would be no useful way in which to have neural mechanisms specialized for particular functions. This is exactly the conclusion O'Keefe and Nadel recognize, and they contest an interpretation resting on equipotentiality as the basis for recovery of function:

> The take-over notion would appear unlikely on purely anatomical grounds. Although one bit of cortex might conceivably substitute for another, the unique machinery of a structure such as the hippocampus is not so easily replaced or simulated. . . . [Most] tasks put to experimental animals can be solved through the use of any of several different strategies, some of which would be dependent upon different brain structures. Recovery of function could thus represent a switch to alternative modes of solution dependent upon intact brain tissue, rather than the actual reorganization of brain tissue. (1978, p. 235)

The denial of recovery of function in favor of a "switch to alternative modes of solution" does much to reinforce the claim to specialization and therefore to localization.[7] To the extent that we see functionally specialized neural systems, we should expect to see localized functions.

3. BOTTOM-UP CONSTRAINTS

The cases discussed in section 2 focus on the analysis of behavioral deficits and the correlation of these with physiological destruction. They are, in fact, reminiscent of the cases discussed in Chapter 4. Nonetheless, they do exhibit marked advantages over the simple localization of Gall, insofar as they introduce more direct behavioral and physiological evidence as well as use pathological data. Their goal is to discern neural correlates of psychological activity. Still, the evidence appealed to in section 2 is still simply correlational: studies have the goal of finding the various mental states and/or behaviors correlated with specific forms of neural functioning. Behavioral deficits are correlated with neural damage. Broca's analysis of "Tan" is, effectively, limited to this stage. There is no independent evidence about the underlying neural structures or their capacities. As we have said repeatedly, this kind of evidence is not sufficiently probative to underwrite any strong conclusions about the character of the neural mechanisms, even though it may be enough to launch a program of research. Correlational studies finally need to be coupled with analytical studies geared to determining how mechanisms consistent with known anatomical and physiological features produce the phenomena we see. It is only once the research program is coupled with these analytical studies that it promises to *explain* the observed correlations and psychological/behavioral capacities; or, at a minimum, to show that it is reasonable to believe there are discrete neural mechanisms responsible for such correlations. Let us see how the cases discussed in section 2 were extended by the introduction of lower-level constraints.

Wernicke on Language

Wernicke settled for the less ambitious goal of showing that it was reasonable to believe there are appropriate neural mechanisms. Arguing for the consistency of the correlational studies with known neuroanatomical facts, he hoped to make it plausible that a mechanism could be found. As we emphasized in section 2.1, Wernicke's psychological decomposition was grounded on an associationistic model of learning and memory, which essentially treated the memory trace as an association, partially defined in terms of sensory-motor associations. Wernicke localized the motor-control region for speech in the anterior lobes (specifically, the first frontal gyrus), and the sensory region in posterior lobes (specifically, the first temporal gyrus). This allowed him to analyze the aphasias as the result of disruptions in normal sensory-motor control.

His choice of loci was in no small part supported by Meynert's physiological taxonomy. Bell and Magendie had shown that, within spinal nerve roots, there was a "specific action" assigned to individual nerve pathways: anterior nerves were specialized for motor functions, and posterior nerves for sensory functions. Meynert generalized the view: Assuming that the function of the cortex was to effect sensory-motor coordination, he projected the basic dichotomy observed by Bell and Magendie to the division between posterior and anterior cortical structures. Meynert distinguished three pathways: the "association pathways" which "begin and end in the cortex" and are responsible for intracortical communication; and two projection pathways which "begin and do not end in the cortex" and are, respectively, responsible for input to and output from the cortical structures. He concluded that sensory functions were to be assigned to the posterior, centripetal, pathways, while motor functions were to be assigned to the anterior, centrifugal, pathways.[8] Wernicke continued to carry the *connectionism* and *associationism*, with their physiological interpretations, into the analysis of the higher functions. The aphasic syndromes would then arise from the disruption of the associations, now conceived as the association pathways within the nervous system.

In the earlier work of 1874, Wernicke recognized three forms of pure aphasias[9]; that is, he recognized three forms of aphasia that exhibited the symptomologies of the pure type. These can be seen by turning to Figure 6.3. A *pure sensory aphasia* (or Wernicke's Aphasia) resulted from the loss of the association between *acoustic representations* of speech (S) and the concept (A), leaving the subject able to talk but unable to comprehend speech. A *pure motor aphasia* (or Broca's Aphasia) resulted from a dissociation of the *motor representations* (M) from ideation (Z), leaving comprehension intact but expression impaired. Finally, a *pure conduction aphasia* would result from a dissociation of the acoustic representations (S) from motor representations (M), leaving both comprehension and expression

Figure 6.3. Wernicke's Scheme for Interpreting the Various Aphasias. Wernicke interpreted the various forms of aphasia in terms of the disruption of specific association tracts. Disrupting the connection between S and A caused a comprehension deficit; disrupting the connection between Z and M produced an expression deficit; and disrupting the connection between S and M resulted in a conduction aphasia. In 1874 these were the only "pure aphasias" recognized by Wernicke, with various mixed aphasias resulting from simultaneous disruption of more than one association tract. Wernicke subsequently recognized more forms of aphasia.

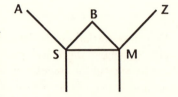

(S) Acoustic Imagery

(M) Motor Imagery

(B) Concept Center

(A) Discharge (Initiation of Conceptualization)

(Z) Ideation (Goal Planning and Motor Integration)

intact, but leaving us unable to mimic heard speech. In terms of Meynert's physiology, the pure conduction aphasias would be the result of disrupting association pathways, while sensory and motor aphasias would be the result of disrupting centripetal and centrifugal pathways.

The general taxonomy of the aphasias was thus at least consistent with the extant neuroanatomical information. This was more than a simple, direct localization, since it proposed a functional decomposition and showed this to be consistent with neuroanatomy and physiology. It remained to be proved that the neural centers were indeed capable of performing the functions assigned to them.

Spatial Memory

O'Keefe and Nadel adopt a more ambitious strategy in supporting localization, arguing that both the functional properties and the anatomical structure of the hippocampus, as well as its place in more general neural structures, are consistent with the hypothesized functions of the hippocampus. They thus argue not only that there is reason to think there are appropriate neural mechanisms, but that the detailed structure of the hippocampus promises to realize the needed functions. Here we find at least two discrete substages in O'Keefe and Nadel's argument: a correlation of psychological activity with detailed neurophysiological functioning, and an explanation of this correlation in terms of the structure of the hippocampus. The first of these substages is still correlational, though carried on at a level of greater detail; the second is no longer merely correlational, but overtly mechanistic.

In the first of these substages, one target phenomenon is hippocampal theta.[10] Theta is a rhythmical, slow activity characteristic of EEG during, for example, exploration, or in the presence of novel stimuli. It serves as a measure of the relative involvement of neural structures. In Figure 6.4, theta activity is especially evident during swimming and the indicated periods of movement. In a summary form, O'Keefe and Nadel (1978) claim that

1. Hippocampal theta is correlated with activity in systems implicated in movement. It is not merely a matter of general arousal or increased motivation (cf. p. 174).
2. Theta is not the result of motor feedback. Peripheral control is ruled out because theta activity can be enhanced even under paralysis (p. 175).
3. Theta must be correlated with some "more global or molar aspect of behavior" (p. 176); it is not limited to the activation of any one specific muscle or muscle group.
4. High-frequency theta is correlated with speed and distance traversed. Looking again at Figure 6.4, the frequency differences in the two periods of movement are correlated directly with these variables (p. 182).
5. Low-frequency theta appears to be correlated with attention and exploration, rather than movement (p. 185); thus, theta presents a complex picture of neural activity *somehow* correlated with movement, but not with *any specific* motor activity.

One of the most salient correlates of hippocampal activity is location. Demonstrating this requires descending to a lower level of analysis, as EEG only reveals more macroscopic phenomena. O'Keefe and Nadel turn

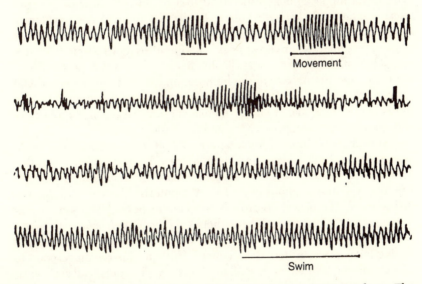

Figure 6.4. A Simplified Illustration of Theta Activity from EEG Readings. The rhythmic, slow activity characteristic of theta is especially evident in the periods of movement and swimming.

to single-unit recordings in the hippocampus. These display a striking correlation between location in a test apparatus and spiking in the appropriate cell. This spiking is limited to specific locations and independent of the type of activity. For example, as a rat reaches a choice-point in a maze, there will be correlated activity in particular cells that is independent of such factors as the direction from which the point is approached. Such correlations are, at least, consistent with the claim that the hippocampus is specialized for some form of spatial memory.[11]

The EEG correlations, and the correlation of hippocampal activity with features that would be required of a local system, are important, but the explanation of these correlations in terms of the structure of the hippocampus is critical to show that we are not engaging in a "vacuous functional localization" (Bechtel 1982)—if we are to show, that is, that the hippocampal structures are the *actual* mechanisms underlying behavior rather than merely potential mechanisms. This takes us to the second substage, emphasizing bottom-up constraints.

Neuroanatomically, the hippocampus has two principal components: the *fascia dentata* and the hippocampus proper (the complete hippocampus is formed by these two interlocking structures, as illustrated in Figure 6.5). Both of these structures are relatively simple. At a cellular level each is fundamentally a single sheet of large neurons with projections in parallel. In the fascia dentata, the primary neurons are granule cells; in the hippocampal regions—CA1, CA2, and CA3—they are pyramidal cells. Within the hippocampus proper (and to a somewhat lesser extent, the fascia dentata), basket cells serve locally as interneurons, evidently subserving an inhibitory effect. The conduction of impulses is, to at least a first approximation, unidirectional, with impulses flowing from the fascia dentata through CA3 to CA2 and, finally, to CA1. The most important afferent tracts (see Figure 6.6) to the hippocampus come from the entorhinal cortex (EC), which exerts a strong excitatory effect on the granule cells of the fascia dentata, and from the medial septum (MS), which projects to cells throughout the hippocampus and the fascia dentata.[12]

Given this structure, O'Keefe and Nadel advance a model of a locale system that is both elegant and straightforward. There are five primary elements, three of which are localized within the hippocampus proper. Inputs to the hippocampus come from the entorhinal cortex and the medial septum. The former controls sensory inputs or configurations of sensory inputs; the latter modulates theta frequency in response to motor control, including an animal's speed or the distance covered. The *initial coding* takes place in the fascia dentata. Units in this region respond to specific sensory inputs, or to configurations of such inputs, and are "sensitive not just to the simultaneous presence of two or more stimuli, but to their occurrence in a unique spatial configuration" (O'Keefe and Nadel

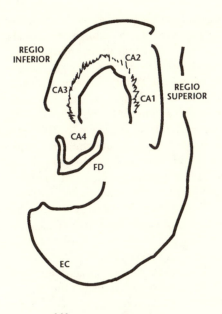

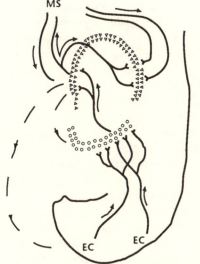

Figure 6.5. A Diagrammatic Representation of Hippocampal Structures. The hippocampus is a region of cortex consisting of two interlocking structures, the fascia dentata (FD) and the hippocampus proper; the latter can be differentiated into three further regions (CA1, CA2, and CA3). It is generally accepted that the hippocampus is critical for encoding of memories, and O'Keefe and Nadel speculate that it is specifically involved in spatial memory.

Figure 6.6. A Second Representation of Hippocampal Projections. This is an extremely simplified representation of projections involving the hippocampus and related structures. The actual connections are far more complex and greater in number than the figure suggests. Inputs come from the entorrinal cortex (EC) and the medial septum (MS), as well as a variety of other extra-hippocampal cortical regions.

1978, p. 221). The output of the fascia dentata is to a *place system* localized in CA2 and CA3. The function of this system is to use the input from the fascia dentata to determine and represent the current place of the animal, and then to represent this in such a way that the current place is located within a system of places having a determinate distance and direction from the current location (cf. O'Keefe and Nadel 1978, p. 223). The place system must tell us where we are, *relative to* other (known) places. For the construction of such a "map," theta rhythms are thought to be crucial. O'Keefe and Nadel conclude (1978, pp. 157ff.) that the medial septum serves as a "pacemaker" designed to impose a synchronous bursting pattern on the hippocampus and the fascia dentata. Within the fascia dentata, this means there will be a pulsed output to CA3 capable of activating only certain pyramidal cells; for example, one can envision a pulse traversing a series of pyramidal cells, activating only the set it reaches at the time that inhibitory processes are minimal. Since theta frequency varies with the rate of spatial displacement, which cells within CA3 are activated will depend, in part, on how rapidly the animal is moving from place to place. A pulse responsive to a particular configuration of stimuli, though projecting to a series of pyramidal cells, will activate only a narrow band. The band activated will then correlate place with output, and relative location on the series will represent external place relative to other locations.

The final component of O'Keefe and Nadel's model is a *misplace system*. It is located, somewhat tentatively, in CA1. Its task is "to generate mismatch signals when the representations of places do not match the stimuli experienced in the corresponding part of the environment" (1978, p. 228). The misplace system consists of units that have higher rates of firing when there is a mismatch: the higher the rate, the more extensive the mismatch. It is not clear to us, in any detail how this mechanism would work.

Appealing to lower levels allows O'Keefe and Nadel to offer an admittedly partial and provisional model of performance based on the organization and function of the hippocampus. The model is no longer merely correlational, but the product of both a higher-level decomposition and a localization constrained by neurophysiological information. In this case, the behavioral features focus on the deficits of hippocampals relative to normals, and the specific pattern of errors of both groups. The resulting functional decomposition is admittedly speculative, but it plays an important role in the program of research. At a finer level of detail, integrating both physiology and behavior, we have correlational information about the relation of behavioral variables to large- and small-scale neural activity. And at a yet finer level we have neuroanatomical and neurophysiological information about the components of the hippocampus, providing evidence that the hippocampus has the computational resources to perform the function assigned to it. While every element of the case, as often is true, is indirect and inferential, that supports no objection whatsoever. What is important for our concerns is that O'Keefe and Nadel supplement the initial behavioral information with evidence garnered at lower levels. The result is a picture of the hippocampus as a complex machine with different parts contributing in characteristic and distinct ways to the overall task. This is localization under multiple independent constraints.

4. CONCLUSION: THE RISE AND DECLINE
   OF DECOMPOSABILITY

The cases described in this chapter can justly claim to have circumvented the central charges against direct localization. As we have seen, opponents of Gall, such as Flourens, were quick to press that his own explanations of behavior were spurious, as they explained behavior in exactly the terms they described it; such explanations were no more than redescriptions. Moreover, they were metaphysical and speculative if they were meant as postulations of occult powers. One central feature of complex localization is that it utilizes a complex strategy in evaluating and developing models; it simultaneously relies on evidence concerning phenomena pitched at different levels of organization. The cases in this chapter provide evidence relevant to psychological processes, and *independent* evidence relevant to neurological functioning. As a consequence, the resulting model is designed to synthesize and accommodate phenomena at more than one level; it acknowledges constraints from two independent directions. Thus, Wernicke's model of language processing depended on a detailed characterization of the aphasic symptomology, because that provided evidence concerning patterns of association. It depended as well on information from Meynert's neurophysiology. Similarly, O'Keefe and Nadel's characterization of the hippocampus as a system for spatial representation relies on behavioral data designed to support a characterization of our capacities and a claim to their independence. It also depends on evidence regarding neurophysiological functioning.[13]

A second feature of complex localization is that, at some point, the researcher does not simply divide up the tasks performed by a system, but rather divides a given task into subtasks that interact in the performance of the overall task. As we will explain in substantially more detail in Chapters 7 and 8, the most natural assumption is that the interactions are linear: one subsystem performs a specific task, and that product serves as the input to the operation at the next stage. In Wernicke's model this decomposition is found in the differentiation of comprehension, production, and association; in O'Keefe and Nadel's work it is found in the differentiation of distinct tasks within the functional organization of the hippocampus. No one task is a primitive or fundamental activity in the sense the phrenologists proposed; each is a complex and composite activity achieved by distinct subsystems realizing their own intrinsic functions. Decomposition of the task, with further decomposition into subtasks in light of localization, is what finally yields a mechanistic explanation.

In the cases discussed in this chapter, the functional decomposition and localization, as it evolved under multiple constraints, maintained much of what a simple localization would have projected. If an initial decomposition and localization do lead to a recognition of one sole, or centrally, responsible component, then, at least as a first approximation, that component can be treated independently. This is what we call *first order independence*. It is one natural outcome of a program of decomposition and localization, and it sets a clear agenda for a research program. We represent this process in Figure 6.7. Figure 6.7 also represents a choice-point in developing a research program: whether to go to a lower level of organization in elaborating the mechanism. When one does integrate a lower level, subcomponents within the system need to be identified and isolated, and their functional properties ascertained. The nature of the interactions of these subcomponents with other systems and components of other systems will evidently also need to be examined, thereby partially compromising the assumption of independence. This is why we refer to the independence as *first order*.

Carrying through this program of research will require operating at more than one level of organization. O'Keefe and Nadel incorporate psychological data, information concerning larger neural structures, as well as a variety of other information, including information about the functional properties of individual cells.

It is in the construction of theories that bridge levels that the importance of decomposition and localization is most clearly manifested. The synthesis of information from more than one level relieves us from methodological limitations of more restricted studies. Though it is often hard enough to find one model adequately capturing the psychological and behavioral data if we are restricted to this data, it is even more difficult to know whether we are confronted with statistical or experimental artifacts. From the other direction, information about neurological functioning in the absence of an interpretive framework can be simply overwhelming. To highlight the point it may help to recall the interpretive principle that brought lesion studies within the compass of O'Keefe and Nadel's purview:

> In the absence of the locale system, hippocampal animals solve problems with taxon hypotheses, and their behavior reflects the properties of the taxon systems. Certain experimental situations will primarily reflect the operations of the taxon systems. . . . In these and other taxon-based situations we would expect the hippocampal animal to perform at least as well as the intact animal. (1978, p. 316)

In their subsequent treatment of operant and classical conditioning, O'Keefe and Nadel explain the lack of a clear *general*—that is, a nonspecific—deficit in hippocampal animals by pressing that the "solution" is available using only taxon hypotheses and according the locale system a "minor" but unspecified role. Many problems, then, can be solved in dif-

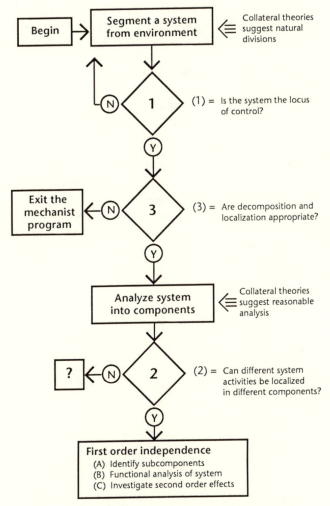

Figure 6.7. One Result of Localization and Decomposition. In some cases a result exhibits first order independence; that is, the primary dimensions involved in system behavior can be explained in terms of the behavior of a single component, in light of a functional analysis of the system. The resulting program of research would extend the analysis by elaborating the mechanisms and investigating the interaction with other components within the system.

ferent ways. In experimental settings we can hope to pull apart the systems and isolate their behavior.[14] This is possible only if the system is either decomposable or nearly decomposable and only by imposing complex constraints.

The critical question is whether there is a decomposition of cognitive functioning that respects neural localization. If there is a failure of localization and decomposition, we can make adjustments at either level, or at both. We could, for example, reject the higher-level, functional characterizations in the face of a failure of localization; that is, failure of localization can be treated as evidence against the higher-level functional analysis. The problem, thus understood, lies in the (potential or actual) mismatch between function and structure. On the other hand, it may turn out that the simple aggregative or linear structure congenial to decomposability and exemplified in the phrenological tradition, as well as the views of Wieland and Warburg on cellular respiration, is untenable; that is, the functional decomposition may fail. These possibilities will occupy us in the next two chapters.

Integrated Mechanisms

1. Introduction: Replacing a Direct Localization

Analysis into localized components and their interactions is a fruitful scientific strategy when the system under study is nearly decomposable; that is, when the organization is relatively simple. In defending *near* decomposability as a heuristic for human problem-solving, Simon offers two markedly different kinds of reasons in its favor. First, given the resource limitations of human beings, near decomposability is an assumption that enables us to deal efficiently with complex systems. This is a kind of naturalistic grounding for simplicity. Second, simply or nearly decomposable systems are more likely to evolve.[1] We have already pointed out that Simon's second reason is less plausible than his first: Evolution works with functioning systems and modifies them to carry out new tasks and meet new demands; it is not creation de novo, but descent with modification. Modifying an existing system may entail altering some of its parts, but this is accomplished by adjusting extant structures to fit new demands. This theory lies at the heart of appeals to evolutionary constraints.[2] As Darwin said, initially appealing to similarities in developmental stages,

> We can clearly understand, on the principle of variations supervening at a rather late embryonic period, and being inherited at a corresponding period, how it is that the embryos of wonderfully different forms should still retain, more or less perfectly, the structure of their common progenitor. . . . Thus we can understand how it has come to pass that man and all other vertebrate animals have been constructed on the same general model. (1871, p. 32)

This development through descent neither depends upon nor creates decomposability. If anything, the mutual coevolution of parts will compromise what decomposability there is and make the system more integrated functionally.

Problems of the sort we will discuss in this and the next chapter present especially difficult problems for localization and decomposition as research strategies because the systems involved exhibit a more complex form of organization. Within some systems, processes depend on the integration of lower-level components, rather than on just their weak interaction. These are *functionally integrated systems*. A research strategy that decomposes a hierarchically integrated system into units exhibiting independence or quasi-independence will not identify the processes that result from the integration of the system.

Simon characterizes the kinds of nearly decomposable systems with which he is concerned as hierarchical, but he explicitly notes that in so doing he is not requiring that hierarchies incorporate subordination of function. This is one important feature of integrated functions in complex systems: they provide a means by which one component can exercise control over other components. To capture the idea that, in some hierarchies, higher levels exercise control over lower levels, Pattee (1973) introduces a distinction between what he calls *structural hierarchies* and *control hierarchies*. Structural hierarchies are Simon's hierarchical systems. In control hierarchies, demands on systemic function affect constituent behavior, and the mode of organization has the consequence that the interactions between components at one level can alter the behavior of the constituents of these components; that is, processes explicable at a higher level alter processes at lower levels.[3] The result, Pattee claims, is that we can no longer identify a sharp cleavage between levels in nature and thereby develop dynamic theories at one level while ignoring or averaging over processes at others:

> In a control hierarchy the upper level exerts a specific, dynamic constraint on the details of the motion at the lower level, so that the fast dynamics of the lower level cannot simply be averaged out. The collection of subunits that forms the upper level in a structural hierarchy now also acts as a constraint on the motions of selected individual subunits. This amounts to a feedback path between levels. Therefore, the physical behavior of a control hierarchy must take into account at least two levels at a time. (Ibid., p. 77)

Control hierarchies, as far as we understand them, result from the way a system is organized. In functionally integrated systems the behavior of the components is interdependent, so a change in the behavior of one part alters the behavior of others. Thus, the *systems* are self-organizing because of the integration and interdependence of component functions.

As with the cases discussed in the previous chapter, the focal case we consider in this chapter—research on fermentation and muscle glycolysis—began with an assumption of simple decomposability that shifted to near decomposability. Fermentation was assumed to involve a series of discrete steps, each providing products serving as precursors for the next. The research problem was then to identify and localize control for each step. This is complex localization in another domain. We will see that research progressed on the assumption of a linear organization.[4] This is a natural extension of localization and decomposition, acknowledging minimal complexity. Linear organization is less complex, and less demanding cognitively, than other forms of organization that might equally well exist. Simplicity is, thus, a virtue of theories, even if only because of our bounded rationality. Unlike the cases of Chapter 6, though, in these instances the assumption was finally the grounds for its own undoing: after attempting for over three decades to develop a model with linear organization, researchers realized that nature employed a far more complex mode of organization.

There is another important difference between the cases. Research in cognitive neuroscience began with direct localization. More complex models followed only with a shift to lower levels of organization, as a natural consequence of the research program. After initially implicating the hippocampus in memory, O'Keefe and Nadel turned to a yet lower level, asking how the components of the hippocampus realized those higher-level functions and what the intrinsic properties of these components were. In our sense of the term (Chapter 2), these researchers follow a largely "analytic" method. The alternative paradigm to be discussed in this chapter arises when localization turns out to be incorrect; instead of one component controlling an activity, a cadre of components produce it jointly. The problem is then not one of isolating the localized mechanisms, but of exhibiting the organization and the constituent functions. The result is an explanation pitched at the same level as the initial localization; an explanation in terms of organization supplants direct localization.

The research program, accordingly, concentrates on articulating the system's organization. Complex localization requires a functional analysis of system behavior, explaining performance as the product of several different and distinct functional systems. The problem then becomes one of identifying component functions and the underlying organization. It is possible to differentiate component parts of the hippocampus physically and to correlate activity in these components empirically with animal behavior; the physical analysis precedes and aids in the functional decomposition. In the cases we will discuss in this chapter, it was only *after* the complex localizationist explanation had been developed that the structures involved in fermentation were identified physically. Enzymes and coenzymes were identified first in terms of the contributions they made to cellular processes—they were functionally defined as whatever facilitated a certain reaction. It was only later that the precise constitution of enzymes could be determined. Likewise, it was only with the introduction of cell fractionation and the electron microscope in the 1940s and 1950s that researchers were able to identify the sites in the cell where critical enzymes were located and synthesized (cf. Bechtel 1989). As a result, the identification of the steps in the fermentation process, and of the enzymes and cofactors that figured in that process, was initially given in terms of the functions these agents served.[5]

The experimental strategy for developing the functional decomposition also differed. For the most part researchers had to forgo analytic approaches and pursue synthetic methods. That is, they employed information about the overall reactions and constructed models of how these might proceed by a sequence of basic chemical processes through known chemical intermediaries. Hence, the first step was to construct a model of a possible chemical pathway. This was constrained primarily by knowledge of net reactions and possible chemical reactions, and thus was speculative. Subsequently, as procedures were developed for studying intermediate stages the experimental task was to find empirical evidence relevant to the theoretically driven models.

We will chronicle some critical elements in the history of research on fermentation and glycolysis from the turn of the century through the 1930s in order to show how researchers managed to push beyond a conception of the cell as decomposable or nearly decomposable and came to recognize an integrated mechanism in cell biochemistry. Researchers at the outset of the twentieth century adopted a simple localizationist view, identifying a single unit within the cell as the agent responsible for glycolysis. This will be discussed in section 2. They soon recognized, however, that the process was more complex, that there were several steps in the transformation of glucose into either lactic acid or alcohol. Once it became apparent that direct localization was inadequate, it was still assumed that the cell was a nearly decomposable system in which different enzymes carry out different operations in a step-wise, linear procedure. Each enzyme was thought to be largely or wholly independent of others, and reactions were catabolic: a sequence of enzymes catalyzed a series of reactions, with the output from each reaction providing the input to the next. This attempt at complex localization will occupy us in section 3. As research continued it became increasingly apparent that even these more complex models could not accommodate the evidence produced through experimental procedures and were incompatible with constraints imposed by collateral theories. During the 1930s it was accepted that this conception of metabolism also failed to account for the complex modes of biochemical organization found in the cell, and that it was this organization which allowed the cell to regulate its own processes and serve its physiological functions. These modes of organization were provided by the substances that, in the linear model, were viewed as simply substances entering or leaving the pathway. But many of these substances in fact connect different reactions in the pathway, providing a kind of architecture in which the whole process occurs. This means that one reaction affects other reactions. These linkages are lost when one assumes that the glycolytic system is nearly decomposable and tries to study individual reactions in isolation. This development of a model with an integrated organization is the emphasis in section 4.

2. DIRECT LOCALIZATION OF FERMENTATION IN ZYMASE

As we have described the case in Chapter 5, during the nineteenth century there was considerable dispute over whether fermentation was due to chemical agents within living cells, or whether it required the whole living cell. This was a dispute over the appropriate level at which explanation should be pitched. Those, like Liebig, who defended a catalytic account of fermentation were committed to holding that the appropriate level of explanation was chemical. Many researchers contended, however, that some reactions could not be carried out by ordinary chemical catalysts, but instead required the entire cell. Processes conceived at the chemical level were, accordingly, inadequate. It was necessary either to introduce novel agents or to appeal to a higher level of organization. In the wake of Schwann's and Pasteur's work, alcoholic fermentation became the chief example of this class. To mark the distinction between those reactions that depended only on chemical agents and those that seemed to require living cells, the purely chemical agents came to be called *unformed* or *unorganized ferments*, while the living cells came to be known as *formed* or *organized ferments*.[6] Kuhne (1877) introduced the term *enzyme* for unorganized, or purely chemical, ferments (see Teich 1981).

The question whether fermentation could be accomplished by an unorganized catalyst or an enzyme, or whether it required a living cell, was actively debated in the third quarter of the nineteenth century. Several prominent, purely chemical theories of fermentation were advanced by investigators such as Traube (1858), Bertholet (1860), and Hoppe-Seyler (1876). These theories did not gain wide acceptance, in part because researchers were unable to isolate from the cell a catalyst or enzyme that could perform fermentation. If an explanation at the chemical level were correct, then it should have been possible to induce fermentation in the absence of living organisms. This, in turn, would be possible were a catalyst isolated and identified.

One of the first attempts to extract a chemical agent capable of catalyzing fermentation was carried out by Ludersdorff (1846). Ludersdorff ground yeast between glass plates. The resulting paste could *not* perform fermentation. In 1872, Marie Manasseïn performed the same test, but claimed to have more positive results (Manasseïn 1897). However, because it was not clear that all the living yeast cells had been destroyed by her grinding techniques, her results were viewed with some skepticism.

Louis Pasteur also engaged in a grinding experiment and failed to find any evidence of a purely chemical agent (see Kohler 1971). Eduard Buchner's (1897) demonstration that fermentation could be accomplished in a yeast press juice from which all living cells and solid cellular material had been removed was therefore quite a surprise and reversed a long train of negative results.

Buchner initially attributed fermentation to a single enzyme he labeled *zymase*. As Kohler (1973b) argues, Buchner's success provided much of the direction taken by biochemistry in the early decades of the twentieth century. The experimental problem was to isolate and identify enzymes responsible for catalyzing each of the reactions in the cell that did not occur spontaneously in conditions like those found naturally in cells. As Kohler says, this problem virtually defined the research program of biochemistry:

> The new profession of biochemistry that began to emerge about 1900 was initially composed of specialists in a variety of established fields, brought together by a common outlook on the physico-chemical nature of life, a common belief that enzymes were the key agents in the life processes, and shared historical experience, and a new name, biochemistry. (Ibid., p. 181)

Buchner's initial model was committed to simple, or direct, localization. This is somewhat surprising in light of the fact that chemists had by that time accepted fermentation as a complex process involving the scissioning of the sugar molecule and the oxidation of part of the resulting material at the expense of other parts. Organic chemists were already attempting to decompose sugar with alkalies to produce alcohol, and they had successfully isolated several intermediary products. A large part of the explanation for Buchner's adherence to direct localization is that he was simply not part of this research tradition.

Buchner's research was actually part of a tradition focusing on immunology. After the development of the germ theory of disease by Pasteur and Koch, the bacteriological community faced an issue parallel to that which had earlier divided researchers working on fermentation. Some researchers conjectured that the agents of bacterial infection might be the chemical constituents of the bacterial cells, and not the cells themselves. Many of the same researchers pursued the idea that the body's defenses against disease were also chemical in nature.

Hans Buchner, Eduard's brother, figured centrally in this controversy, but took a middle ground between chemists and organicists. He argued that the antibacterial agent was the "living blood plasma," which he took to be a *living* protein (H. Buchner 1889). The idea of living protein was part of a *protoplasm theory* adopted by a number of researchers who tried to integrate chemistry and physiology in the late nineteenth century by arguing that the basic material of living organisms was chemical in nature, but that this material assumed a special form with special properties in living organisms.[7] Unlike most chemically oriented researchers, who focused on chemical materials as they might be excreted from the cell, Hans Buchner was interested primarily in isolating the proteins that resided *in* the cell.

The project that linked the work of Hans and Eduard Buchner was devoted to grinding bacterial cells to remove the intracellular proteins. To do this Eduard developed a technique for grinding with sand. The debris, however, contained remnants of the cell bodies, and for that reason was regarded as unreliable. It was not until 1896 that one of Hans Buchner's associates, Martin Hahn, developed a technique for filtering the debris to produce the cell juice. This juice rapidly decayed, and it was in the process of trying to find a preservative that sugar was added to it. Eduard Buchner recognized the resulting reaction as fermentation. This was the first clear case of fermentation in a pure cell extract (see Kohler 1971). Because the critical work came out of a project that was principally devoted to isolating bacterial toxins and antitoxins, the interpretation of the work as finding *the* chemical agent responsible for fermentation was not surprising. Within his own problem domain, Buchner's search was clearly directed at whether it was possible to isolate single chemical factors that would produce the pathological reaction. Because the cell extract did produce fermentation, it was natural to conclude that the same agent—whatever it might be—was responsible in intact cells. Moreover, the program of research suggested the agent was a *single* chemical agent.

Eduard Buchner's chemical interests led him to look further at the process by which sugar could be rendered into carbon dioxide and alcohol. By 1904 he provided evidence that "lactic acid plays an important role in the cleavage of sugar and probably appears as an intermediate in alcoholic fermentation" (Buchner and Meisenheimer 1904, pp. 420–21). Eventually Buchner abandoned the simple localization of fermentation in zymase and came to view fermentation as at least a two-step process: Zymase was to be responsible only for the first reaction, producing lactic acid from sugar. The reduction of lactic acid to alcohol was mediated by what he called *lactacidase*.

At the time of the Buchner's success in producing cell-free alcoholic fermentation, the production of lactic acid from sugar in the course of muscle work also was classified as a type of fermentation.[8] The initial linkage of the production of lactic acid with muscle work stemmed from du Bois-Reymond's (1859) discovery of lactic acid in muscles after muscle contraction or the death of the animal. Hermann (1867) inferred that this reaction was anaerobic, as is alcoholic fermentation, and Bernard showed that it was a general reaction in animal tissue:

This lactic ferment occurs in the blood, in the muscles, even in the liver, since I have found that muscle and animal tissues do not become acid after death unless they contain sugar or glycogen which rapidly undergoes a lactic fermentation. (1877a, p. 328)

The final convincing demonstration of the linkage between lactic fermentation and muscle work was found in the research of Fletcher and Hopkins (1907). As we shall see, the relation between alcoholic fermentation and lactic acid fermentation was crucial in unraveling the mechanism of fermentation. Eventually biochemists came to use *glycolysis* to refer generically to both processes.

3. A COMPLEX LINEAR MODEL OF FERMENTATION

The demonstration that fermentation could be carried out independently of living cells reawakened interest in developing a chemical account of fermentation. The task of determining the sequence of reactions fell to biochemists during the first three decades of this century and culminated in the basic understanding of the chemical stages in fermentation developed in the 1930s. Biochemistry placed several constraints on an acceptable solution to the problem: the first is an operational constraint, requiring independent evidence for the processes involved; the second located the processes at the chemical level; and the third is tantamount to near decomposability. Let us look briefly at each and then turn to consider their impact in more detail.

The requirement of independent evidence for postulated processes is particularly challenging in the case of reactions within living systems, because it is difficult to gain direct access to the reactions. As in many other cases, researchers of fermentation had to rely on indirect evidence. It was assumed that fermentation began with the scissioning of the sugar molecule into three-carbon sugars and ended with the decarboxylation of a three-carbon compound to form alcohol. The problem was in determining the intervening reactions. It was not enough to produce a chemical model of *possible* intermediaries; it was necessary to determine the pathway *actually used* in living organisms. One way to demonstrate the correctness of a model pathway was to show that the proposed intermediaries actually did arise in the process of fermentation.

One source of evidence for the occurrence of a proposed intermediary would be the discovery of a small quantity of a substance among the reaction products. This proved to be difficult technically, mainly because metabolic reactions tend to occur rapidly, with the intermediate products from one reaction being consumed almost immediately in subsequent reactions. Another, more useful source of evidence was an increase in concentrations of a potential intermediate when the reaction was interrupted in some way. In classical inhibition studies of this type, a variety of substances were used to poison the cell, stopping the reaction at a specific stage and bringing about the buildup of intermediate products. Evidence that a potential intermediary could be metabolized by the living tissue in question served as a third source of information. One of the principal tests used to determine which of the intermediaries actually figured in fermentation was to insert the intermediary into cells, or (more often) cell extracts, to determine whether it was consumed in the reaction. Researchers generally described these three procedures as tests of whether the intermediary would itself "ferment."[9] It was important not only that the potential intermediary ferment, but that its rate of fermentation be at least as rapid as that of the sugar itself. Failure to ferment at an acceptable rate would count against the substance being an actual intermediary in fermentation and, accordingly, against the model.

The second constraint on an acceptable explanation of fermentation was that *each* stage in the process must involve a known, basic chemical reaction. From a chemical perspective, it had long been clear that fermentation was a complex process. As organic chemists understood reactions toward the end of the nineteenth century, they held that basic chemical reactions involved the addition or removal of small chemical groups (see Table 7.1). These constituted the reactions nineteenth-century biochemists were prepared to accept in models in which fermentation was regarded as a chemical process. The fact that fermentation was obviously not such a simple chemical reaction, but was far more complex, placed it outside the domain of basic chemistry. Those researchers with a chemical orientation recognized the need to break the reaction into a sequence of reactions, where the component reactions would be the required, basic chemical reactions.

The third constraint was the assumption that the overall reaction involved a linear degrading of the initial substrate to produce the output. This assumption was tantamount to near decomposability because, if it held, each reaction necessarily occurred in relative independence and

Oxidation	The addition of an oxygen molecule.
Reduction	The addition of a hydrogen molecule.
Dehydration	The removal of a water molecule from substrates.
Decarboxylation	The removal of a carbon dioxide molecule.
Deamination	The removal of an ammonia molecule.

Table 7.1. Basic Chemical Reactions. These were some of the allowed chemical reactions as understood by organic chemists toward the end of the nineteenth century.

could be studied in isolation, provided the appropriate inputs were available. The idea that fermentation would be linear was natural at the time. On the one hand, the product of fermentation—alcohol—was far simpler in structure than sugar. On the other hand, the overall reaction released energy. Research then progressed on the assumption that fermentation was a sequence of catabolic reactions, with each step yielding a simpler intermediary and releasing energy until alcohol was produced.[10]

While there was explicit reference to the first two constraints, this last one was generally left implicit. Knoop (1904) came close to an explicit statement of it, though, after proposing the model of β-oxidation for fatty acids. He said the goal was "a knowledge of the course of the decompositions and oxidations in the building materials and nutritional substance of the animal organism that would leave no gaps" (ibid., p. 3). By referring only to decompositions and oxidations, Knoop left no place for syntheses and reductions, and hence appeared to be calling for a linear model. In discussing the linearity of metabolic mechanisms, Thunberg was more explicit:

> I consider that nothing of the oxygen consumed in the general metabolism is found in the expired carbon dioxide. I consider the catabolism of the food stuffs to take place in a series of continuous dehydrogenations, carried out by a series of dehydrogenases. This procedure, to which the complicated food-stuff molecules are thus subjected, might be compared to what happens in modern factories where a piece of metal glides along on rails from workman to workman each of whom has his special task to carry out in the course of the work until the metal piece leaves their hands as a finished product. (1930, p. 327)

In one sense linearity may have represented an attempt to start with the simplest model and to introduce complications only as needed. However, a linear model was in fact a logical presupposition of the requirement that proposed intermediaries ferment as rapidly as sugar. If the pathway were not linear, then skipping an earlier step might impede later steps and so undercut the assumption that the reaction of an intermediary could proceed without starting at the beginning of the reaction pathway. Linearity was deeply ingrained in the biochemical research program.

Each of these assumptions can be more fully illustrated in work within the biochemistry of the period. We will take them up in turn and consider their implications.

The Search for Intermediates

After Buchner's success in cell-free fermentation, investigators began to search for evidence as to which organic substances might figure in biological fermentation. As we have noted, Buchner proposed a two-step process when he modified his initial, simple localizationist account of fermentation, using lactic acid as an intermediary. Not only had lactic acid been generated as an intermediary in attempts to degrade sugar with alkalies, but Buchner and Meisenheimer (1904) claimed to have found small amounts of lactic acid as products in alcoholic fermentation. Lactic acid was, however, soon rejected as an intermediary by most researchers. The crucial result was Slator's (1906) demonstration that lactic acid could *not* be fermented at all rapidly by yeast.[11] He concluded on that basis that it did not figure importantly in the pathway.

An independent line of research provided evidence of other intermediaries. In the late nineteenth century, organic chemists had carried out a number of experiments, treating glucose with different alkalies. Strong alkalies such as caustic potash resulted in the production of alcohol and carbon dioxide, while weaker alkalies produced lactic acid (Duclaux 1896). Further research in this vein suggested that several three-carbon compounds, including methylglyoxal, glyceraldehyde, and dihydroxyacetone, might be intermediaries in the process (for an overview, see Harden 1932). Attention then turned to these three-carbon substances and, in particular, to whether these substances would ferment as rapidly as sugar. Some researchers, including Buchner, found evidence that methylglyoxal was not able to be fermented by yeast. Nonetheless, as we shall see, it figured centrally in one of the first comprehensive models of fermentation. Evidence was also presented showing that dihydroxyacetone and glyceraldehyde did ferment in living yeast, although the evidence was controversial.

Otto Neubauer's research on amino acid metabolism drew attention to pyruvic acid as another potential intermediary. Neubauer and Fromherz were investigating the oxidative deamination of alanine in yeast and showed that the reaction did not stop with pyruvic acid, but yielded alcohol. Having established the role of pyruvic acid as an intermediary in alanine metabolism, Neubauer proposed that it might also play a role in sugar fermentation. He said pyruvic acid

> could be an intermediate in the alcoholic fermentation of sugar. . . . We ask colleagues to leave to us the further study of the role of pyruvic acid in the fermentation of sugar; also, it is intended to study the question whether it is an intermediate in the combustion of sugar in the higher animal organism. (Neubauer and Fromherz 1911, p. 350)

Evidence rapidly accumulated supporting pyruvic acid as an intermediary. Neubauer demonstrated that it could be fermented, and Carl Neuberg demonstrated the same for pyruvate. Neuberg also identified an enzyme, carboxylase, which catalyzed the decarboxylation of pyruvic acid to acetaldehyde and carbon dioxide. And Fernbach and Schoen (1913) finally isolated a calcium salt of pyruvic acid in the products of fermentation carried out in the presence of calcium carbonate.

Evidence for the occurrence, or at least the potential reactivity, of intermediates thus played an important operational role. In the case of Buchner, the observation that lactic acid could not be fermented naturally served as a reason for rejecting his proposed model. In the case of Neuberg, the ability to naturally degrade pyruvic acid provided a defense for the feasibility of his model. As we shall see, however, this evidence concerning the existence and reactivity of intermediaries provided only a relatively weak constraint on the acceptability of a model of fermentation.

The Appeal to Basic Chemical Reactions

The research on the decomposition of sugars by alkalies resulted in numerous models in which basic chemical reactions accounted for each proposed stepwise change. These models, in turn, provided inspiration for developing models of biological fermentation. Although a number of proposals were introduced in the first decade of the twentieth century, the first to command wide acceptance was developed by Carl Neuberg (Neuberg and Kerb 1914). Neuberg thought the glucose molecule was separated into two molecules (see Figure 7.1) of methylglyoxal (step 1), one of which was reduced to glycerol, yielding pyruvic acid as a byproduct (step 2a). The pyruvic acid was then decarboxylated, yielding acetaldehyde and carbon dioxide (step 3). The acetaldehyde produced in this step participated in a subsequent oxidation, being reduced to alcohol while another molecule of methylglyoxal was oxidized to pyruvic acid. After the initial release of pyruvic acid, step 2a was unnecessary, with the reaction proceeding through step 2b. Neuberg's model was widely accepted during the next twenty years. It had many virtues. Besides providing a sequence of reactions that produced alcohol from sugar through simple known reac-

---

(1) $C_6H_{12}O_6$ $\Rightarrow$ $2\,C_3H_4O_2$ + $2H_2O$
[Hexose] [Methylglyoxal + Water]

(2a) $2C_3H_4O_2 + 2H_2O$ $\Rightarrow$ $C_3H_8O_3$ + $C_3H_4O_3$
[Methylglyoxal + Water] [glycerol + pyruvic acid]

(2b) $C_3H_4O_2 + C_2H_4O + H_2O$ $\Rightarrow$ $C_3H_4O_3$ + C_2H_5OH
[Methylglyoxal + Aldehyde + Water] [Pyruvic Acid + Alcohol]

(3) $C_3H_4O_3$ $\Rightarrow$ C_2H_4O + CO_2
[Pyruvic Acid] [Acetaldehyde + Carbon Dioxide]

Figure 7.1. Glycolytic Pathway Proposed by Neuberg and Kerb (1914). Neuberg and Kerb developed a simple model explaining the breakdown of glucose into alcohol in fermentation.

tions, his model could account for the methyl that had been found among the products of fermentation.[12]

There were also important objections to Neuberg's model, two of which are of particular note for our purposes. The first concerned the role of phosphates in fermentation. Although Buchner showed that fermentation could occur in cell-free extracts, it occurred much more slowly than in living cells. Moreover, after a short period the reaction almost totally stopped. Arthur Harden (1903) established that adding blood serum would yield an 80% increase in fermentation. In subsequent collaborative work with William Young (Harden and Young 1906), Harden showed that adding a phosphate could maintain the reaction. Harden and Young (1908) proposed that this phosphate, which appeared to be consumed in the reaction, was taken up into an ester—hexosediphosphate—with one molecule of the sugar, while a second molecule was fermented. The resulting hexosediphosphate slowly decomposed in a hydrolytic reaction, liberating sugar and phosphate which could again enter the glycolytic reaction (see Figure 7.2). What is especially noteworthy is that Neuberg's model did not provide a role for phosphates, though they did appear to be integral to the process. Several researchers faulted the model on this point, but Neuberg dismissed the evidence for incorporating phosphates, as well as for the production and accumulation of hexosediphosphate, in yeast-juice preparations as an artifact of the experimental design. He contended flatly that phosphates could not figure in fermentation in living cells, because hexosediphosphate was not fermented in living cells (Neuberg and Kobel 1925).

Neuberg's objection to phosphate models rested on the principle that nothing could be an intermediary in fermentation that could not be fermented *as quickly as* the sugar substrate—a principle that seemed entirely reasonable on the surface. If we assume that the overall process is nearly decomposable into its component steps, each of which can be performed in isolation, then a substance that cannot undergo the necessary reaction cannot be an intermediary. Given this much, the rejection of

---

Glycolytic Reaction:

$$2C_6H_{12}O_6 + 2R_2HPO_4 \Rightarrow C_6H_{10}O_4(PO_4R_2)_2 + 2C_2H_5OH + 2CO_2 + 2H_2O$$
[Hexose + phosphate]   [Hexosediphosphate + Alcohol]

Hydrolytic Reaction:

$$C_6H_{10}O_4(PO_4R_2)_2 + H_2O \Rightarrow 2C_6H_{12}O_6 + 2R_2HPO_4$$
[Hexosediphosphate]   [Hexose + Phosphate]

Figure 7.2. Harden and Young's (1906) Conception of the Role of Phosphates in Glycolysis. An alternative model of glycolysis, depending critically on phosphates.

hexosediphosphate as an intermediary in fermentation seems quite sensible.

Ironically, Neuberg's own model was subject to criticism on this same point. Despite numerous attempts, researchers could not show that methylglyoxal could be fermented in living yeast. In this case, however, Neuberg responded quite differently, proposing ad hoc explanations for this failure. He suggested, for example, that the added methylglyoxal might have a different structure than that produced naturally in the reaction and that it might not be able to reach the necessary reaction site within the cell.

The Assumption of Linearity

We have already seen that linearity had an effect on the work on fermentation. The hypothesis that intermediaries must ferment as rapidly as sugar itself assumes that later stages in the reaction are dependent on earlier stages only for their products. In fact, the one intermediary considered that actually does figure in fermentation was ruled out on just this ground: hexosediphosphate accumulated in the course of fermentation in cell extracts in part because reactions later in the normal process were cut short. The result was that reactions earlier in the pathway were also halted.

We have noted that linearity is a natural simplifying assumption. Chains of reactions are more readily tracked and understood than ones involving cycles. However, like other constraints, this assumption can be abandoned when other considerations tell against it, or when it conflicts with the other constraints we have mentioned. This can be seen in failures of linearity in some early models, such as Neuberg's. The final product of the reaction in Neuberg's model, ethyl alcohol, was in a lower oxidation state—that is, it had a higher energy level—than pyruvic acid, which was supposed to be an intermediate product in the reaction. This meant that the formation of pyruvic acid required an endothermic reaction and thus violated the assumption that the process involved simply a sequence of catabolic reactions. Moreover, Neuberg's model assumed that the aldehyde formed from pyruvic acid figured in an *earlier* reaction in which it was reduced at the expense of methylglyoxal (and which was in turn oxidized to yield more pyruvic acid). Effectively, a feedback loop was being inserted into the overall pathway. Not only was step 3 dependent on step 2b, but 2b was dependent on 3. The feedback loop was included because the alternate course, represented by step 2a, would lead to a buildup of glycerol. Glycerol was not found in normal cells, and no alternative set of reactions could be conceived through which the glycerol would be broken down. Our acceptance of this willingness to compromise linearity is no more than a recognition that problem solving is a process of simultaneous constraint satisfaction.

4. An Integrated System Responsible for Fermentation

Gustav Embden and Otto Meyerhof were pivotal in overcoming the assumption of linearity and establishing the integrated character of cellular fermentation. Neither initially rejected a linear organization, but the synthetic character of their approaches eventually led them to develop nonlinear models. One distinctive feature of their methodology was that they did not limit their focus to the basic chemical information and biochemical modes of experimentation. Each conceived of the problems physiologically, and therefore functionally, and each developed models of the physiological role of fermentation. As a result they looked at fermentation in the broader context of cell life and not the narrower context of an isolated biochemical problem. They fostered the development of models that more obviously violated the assumption of linearity and which therefore had the effect of undermining near decomposability.

Embden and Meyerhof: The Role of Phosphates

Embden introduced the idea of phosphorylated intermediaries in the course of research into parallels between alcoholic fermentation and lactic acid fermentation in muscle glycolysis. This in turn was part of an attempt to explain how lactic acid fermentation figured in producing energy for muscle contraction. Embden and his collaborators attempted to simulate Buchner's work with alcoholic fermentation by carrying out lactic acid fermentation in muscle extract. They established that lactic acid increased in the muscle extract over time, but noted that adding glucose failed to increase the yield (Embden, Kalberlah, and Engel 1912). Embden hypothesized that the lactic acid was derived not directly from glucose but from an unknown precursor which he called *lactacidogen*. Embden, noting that lactic and phosphoric acids were produced in his extract in equimolar proportions, concluded that lactacidogen was the precursor of both substances. Because lactacidogen was likely to be a phosphorylated sugar, Embden decided to test the hexosediphosphate ester of Harden and Young by adding it to his muscle extract. That too increased the production of lactic acid. As a result, Embden proposed that it might be identified with lactacidogen (Embden, Griesbach, and Schmitz 1914).

Meyerhof also considered the possibility that phosphorylated substances might be intermediaries in fermentation. In 1924 he offered evidence that phosphorylated hexosediphosphate split into triose diphosphates without first undergoing hydrolysis as Harden and Young proposed. Subsequently he provided further evidence for the role of phosphates by using fluoride to inhibit glycolysis. This yielded a buildup of hexosediphosphate and indicated a more active role for this substance than had been envisioned by Harden and Young. Meyerhof, however, confronted the same obstacle others faced: hexosediphosphate could not be fermented in artificial solutions. In response he proposed that a special, active form of hexosephosphate served as an intermediary. One molecule of this active form would pass its phosphate to the other, creating hexosediphosphate as "a stabilization stage," while the other split into two trioses. The hexosediphosphate itself would subsequently split into two monophosphates that could reenter the pathway.[13]

In the early 1930s Embden took a step beyond Meyerhof, proposing a model in which phosphorylated substances were the precursors of lactic acid. According to this model (Figure 7.3), hexosediphosphate is scissioned into two triosephosphates. One of these triosephosphates is oxidized to 3-phosphoglyceric acid while the other is reduced to glycerophosphoric acid. The 3-phosphoglyceric acid is next dephosphorylated to pyruvic acid and then reduced to lactic acid. At the same time, the glycerophosphoric acid is oxidized to form triosephosphate (Embden, Deuticke, and Kraft 1933). Embden's model was not entirely a speculative proposal designed to allow for a phosphorylated precursor to lactic acid. There was empirical evidence for the critical first step in earlier work by Lipmann and Lohmann (1930). However, when Embden published his scheme the proof for many of its parts was not yet developed. The model eventually proved incorrect in detail, but the development of a model using phosphorylated intermediaries to account for physiological phenomena was a critical ingredient in ultimately coming up with more adequate theories. Embden's awareness that muscle contraction produced free phosphate put him on the path to a radically different understanding of fermentation and facilitated the discovery that phosphate bonds provided the mechanism by which the energy released in the metabolic reactions was transferred to points where it was used.

Acceptance of the importance of phosphate bonds to fermentation was somewhat impeded by the assumption that fermentation must be a series of catabolic reactions; that is, by the assumption of linearity. As a result of the work of Fletcher and Hopkins (1907) it had been generally assumed that the energy liberated in lactic acid formation was the direct source of the energy for muscle contraction, with the energy being transferred in the form of heat. Fletcher and Hopkins's investigations inspired a number of detailed quantitative studies of the relationship between glucose levels, lactic acid levels, and the response of each to oxygen. These studies were coupled with investigations, carried out principally by Hill and Meyerhof, into the relation between the chemical events in the cell and the heat production. Three important results emerged that were regarded as anomalous on the assumption that lactic acid formation was the direct source of energy for muscle contraction. First, heat production was

$$C_6H_{10}O_4(PO_4H_2)_2 \Rightarrow 2C_3H_5O_2(PO_4H_2)$$
[hexosediphosphate]   [triosephosphate]

$$2C_3H_5O_2(PO_4H_2) + H_2O \Rightarrow C_3H_5O_3(PO_4H_2) + C_3H_7O_2(PO_4H_2)$$
[triosephosphate]   [3-phosphoglyceric acid + glycerophosphate]

$$C_3H_5O_3(PO_4H_2) \Rightarrow C_3H_4O_3 + H_3PO_4$$
[3-phosphoglyceric acid]   [pyruvic acid + phosphate]

$$C_3H_4O_3 + C_3H_7O_2(PO_4H_2) \Rightarrow C_3H_6O_3 + C_3H_5O_2(PO_4H_2)$$
[pyruvic acid + glycerophosphate]   [lactic acid + trisephosphate]

Figure 7.3. Glycolytic Pathway Proposed by Embden, Deuticke, and Kraft (1933). This model incorporated phospholated intermediaries as precursors to lactic acid.

greater during oxidative recovery (when lactic acid was metabolized) than during anaerobic contraction (when lactic acid was formed). This was puzzling because the greatest energy expenditure was during the anaerobic phase. Second, the heat produced during contraction was greater than that theoretically available in the production of lactic acid from glycogen. And third, the heat produced during recovery was less than would be expected on the basis of complete oxidation of lactic acid. These discrepancies made it difficult to account for the energy of muscle contraction directly in terms of the thermodynamics of metabolism (see Needham 1971).

Meyerhof's proposal of a lactic acid *cycle* provided the first apparently adequate explanation of these results. Whereas earlier accounts construed all the lactic acid as being oxidized during recovery, Meyerhof (1924) proposed that approximately three-fourths of the lactic acid produced in anaerobic contraction was reconstituted as glycogen during recovery. He postulated that the energy for resynthesis came from coupling the resynthesis with the oxidation of the smaller portion of lactic acid. The fact that not all lactic acid was oxidized would explain why the heat production in this phase was less than expected.

The incorporation of this cycle into Meyerhof's model was important for a number of reasons. It represents one of the earlier cases in which a more complex mode of organization was posited, rather than a simple catabolic process (Nachmansohn 1972, p. 5). It was also important that a coupling of reactions—a nonlinear dependence—was postulated simply to handle the thermodynamics, even though Meyerhof lacked any knowledge of the mechanism for coupling. The constraints that led to a more complex organization were top-down rather than bottom-up. The lactic acid cycle, however, could not account for the additional heat produced during anaerobic contraction. Fitting the chemical events with the thermodynamic data required the recognition that the energy for muscle contraction did not come directly from glycolysis. Embden offered evidence as early as

1924 that not all the lactic acid formation occurred in the course of muscle contraction, but making systematic sense of this depended on finding some other source for the energy of contraction. The solution was found only with the identification of the function of two other compounds discovered in the late 1920s. Eggleton and Eggleton (1927) isolated a rapidly hydrolyzable substance from the cell that released large quantities of phosphorus. At the time this substance was called *phosphagen*, but is now known as *phosphocreatine*. Lohmann (1929) and Fiske and Subbarow (1929) isolated another compound that was readily hydrolyzed into adenylic acid and pyrophosphate. This substance, first named *andenylphyrophosphate*, subsequently came to be known as *adenosine triphosphate* or ATP.

The ATP/ADP Cycle

Once ATP and phosphocreatine were chemically identified as cellular substances, the observation was made that both of these, upon hydrolysis, released large quantities of heat (Meyerhof and Lohmann 1927). The recognition that these compounds were indeed the proximate agents of energy for muscle contraction and the way in which the compounds functioned resulted from an inhibitory study by Lundsgaard (1932). He injected rabbits with iodoacetate and discovered that it was highly toxic. The pattern of this toxicity was important. With exercise the animal entered rigor mortis without the formation of lactic acid. Lundsgaard also showed in extract from frog muscles poisoned with iodoacetate that muscular contractions prior to the onset of rigor mortis occurred without the formation of lactic acid, but with the breakdown of phosphocreatine. This established that phosphocreatine could provide the energy for muscular contraction and suggested that it was also the source of energy in normal muscle.

The scene was now set for Meyerhof and his coworkers to bring ATP into the account of glycolysis and to piece together a picture of the previously mysterious role of phosphate bonds in glycolysis. Lohmann (1931) established that ATP was necessary for fermentation, and, together with Meyerhof, he suggested that ATP served as a link between glycolysis and the resynthesis of phosphocreatine:

> The adenylpyrophosphate [or ATP] cycle maintains the lactic acid formation. The synthesis of phosphagen [or phosphocreatine] is therefore made possible . . . by the cleavage energy of the adenylpyrophosphate, while the energy of lactic acid formation (from phosphate esters) serves to resynthesize the cleaved pyrophosphate. (Meyerhof and Lohmann 1932, p. 576)

Lactic acid fermentation was linked to other processes through which ATP and phosphocreatine were synthesized and broken down. It was no longer viewed as a discrete process.

Meyerhof, Lohmann, and Meyer (1931) recognized that ATP was a critical component in the coenzymes of both alcoholic and lactic acid fermentation. The need for a coenzyme had previously been seen in the same investigations of Hardin and Young that showed the need for a phosphate. Since the coenzyme was needed for the early steps of glycolysis, this suggested that ATP figured there, too (it contributes one of the phosphate bonds in the formation of hexosediphosphate). Meyerhof and Lohmann (1932) established, further, that the complete hydrolysis of ATP (to AMP) released sufficient energy to synthesize two molecules of phosphocreatine, while glycolysis was capable of bringing about the resynthesis of ATP. They pieced these clues together to propose an integrated system in which phosphates would provide the linkage through several steps in the energetic process:

> The present experiments lay the foundation of the thesis that the endothermic synthesis of phosphocreatine can take place through a coupling of this process with the exothermic and spontaneous breakdown of ATP, whilst the resynthesis of ATP out of adenylic acid and inorganic phosphate is made possible through the energy of lactic acid formation. One may also assume here a coupling of the synthesis with the metabolism of the intermediate hexose esters and so see in the phosphate groups contained in all these compounds, the unique carriers of the chemical coupling process. (Ibid., p. 460)

This research made it clear that ATP served an important function in integrating biochemical processes. Further investigations by Lohmann established that ATP also functioned in the hydrolysis of creatine phosphate during muscle action. He also suggested that ATP regulated glycolysis, insuring that it occurred when needed to replenish energy used:

> Viewed teleologically, this dual function seems to be a very ingenious arrangement for insuring the orderly sequence of the chemical processes involved in the muscle twitch. . . . The contraction brings about a fission of adenylpyrophosphoric acid which in turn imposes a cleavage of creatine phosphate, thereby simultaneously reconstituting adenylpyrophosphoric acid; the latter can now interact as co-enzyme by mobilizing glycogen for lactic acid formation. (Lohmann 1934/1969, p. 60)

Another crucial link in the process of ATP formation was recognized in 1937, when Needham, in a theoretical paper reminiscent of those of Embden and Meyerhof, argued for the existence of a second process for esterification of phosphate. She noted that fluoride would stop ATP formation, given the Meyerhof and Lohmann model. She continued to point out that in both yeast and muscle extract poisoned with floride, some ATP does form. Needham also argued that empirical studies showed (1) that more creatine phosphate formed per molecule of lactic acid than the transfer from phosphopyruvic acid to ATP would account for and (2) that the heat output was less than predicted. She therefore proposed that the reaction was accompanied by a second synthesis of ATP. This prediction was confirmed by Needham and Pillai (1937), who established the coupling of the oxidation-reduction reaction and ATP formation.

By the end of the decade a coherent picture emerged of the role of phosphates in glycolysis and in providing energy for muscle action. Glycolysis began with the phosphorylation of glucose, partly at the expense of ATP, to form hexosediphosphate. This was scissioned into two triosephosphates, which were then further phosphorylated and oxidized. The phosphate bonds in this oxidized product now had a high heat of hydrolysis, which was carried over when the phosphates were transferred to ATP and phosphocreatine. The hydrolysis of these substances provided a source of energy for muscle work.

The circulation of phosphates provides a biochemical integration to the system. Whereas glycolysis has previously been assumed to be a linear and nearly decomposable process, with a series of discrete catabolic reactions, the model that emerged in the 1930s revealed the system as highly integrated. If we focus on the breakdown of sugar to alcohol or lactic acid as central, the process can be represented as a linear pathway (see Figure 7.4). The coenzymes that link cellular reactions are simply shown as byproducts peripheral to the main process. This ignores the role of these coenzymes in integrating cellular reactions. These substances—ATP, ADP, NADH, and NAD$^+$—function as links in a complex process that is more perspicuously represented as cyclic (Figure 7.5). What constitutes a product of one reaction is an input to other reactions that occur earlier in the breakdown of glucose to lactic acid or alcohol. These linkages make the fermentation system one that cannot be simply decomposed into its component reactions without seriously misrepresenting the processes occurring in the cell[14] (for further development of this case, see Bechtel 1986b).

5. Conclusion: The Discovery of Integration

We have focused on the process by which biochemists came to unravel the mechanism underlying fermentation. Buchner, in the wake of his discovery of cell-free fermentation, defended direct localization, assigning the reaction to a single enzyme which he labeled zymase. Had it turned out

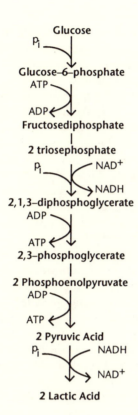

Figure 7.4. A Standard Representation of Glycolysis as a Linear Process. The breakdown of glucose is represented as a linear process with a variety of intermediaries and several byproducts, including ADP, ATP, NADH, and NAD^+.

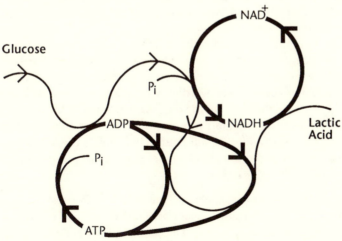

Figure 7.5. An Alternative Representation of Glycolysis as Cyclic. This alternative (and somewhat simplified) representation acknowledges the critical role of the ATP-ADP cycle and the NADH-NAD^+ conversions in glycolysis. Linearity is explicitly compromised.

that one enzyme was responsible for fermentation, research should then have turned to a lower level to explain how this enzyme functioned. In this case, however, direct localization was flawed. Fermentation is not a simple chemical reaction; a number of enzymes and coenzymes figure in the process. An explanation at the same level as the incorrect direct localization was required to understand fermentation.[15] We have identified a variety of constraints on the emerging explanation. It was, at least initially, assumed that the overall process would be decomposable into a linear sequence of reactions. The product of one reaction would serve as the substrate for the next, and the ordering would be provided by the energy available in the compounds. This assumption was coupled with the hypothesis that any intermediary in the process must ferment as rapidly as the sugar itself. These assumptions led researchers to dismiss phosphorylated substances from the pathway for nearly twenty years and to search instead for a means by which methylglyoxal could be fermented. Ultimately it was discovered that the fermentation system was minimally decomposable. Various reactions were mutually interdependent. If we experimentally disrupted these reactions, then even a normally fermentable substance would not ferment. Even more significantly, disrupting one step in the pathway would block others that preceded it. We now think of the breakdown of ATP to ADP as the critical reaction, since it releases energy for work. However, disrupting processes that would otherwise appear obviously downstream (for example, the formation of pyruvic acid) will inhibit the formation of ATP. Isolating the process of glycolysis from that in which ATP is broken down to ADP also prevents glycolysis, as some of the reactions in glycolysis require ADP. Interaction and organization in the system are critical.

Discovering the organized context within the cell permitted researchers to overcome the long-standing debates over whether processes in the living cell are common chemical effects or of some different nature. The apparent differences between ordinary chemical reactions and those in living systems are the result of the organization found in the cellular environment. The basic reactions are chemical, and organization serves to modulate their operation. Rudolf Peters (1957) offers the following retrospective account of the transformation of biochemistry as it came to deal successfully with organization:

> We seem to have travelled very far from the old controversies, now hardly embers even. That of vitalism versus mechanism seems to me particularly dead at the moment, because most of us, I imagine, now feel that the one thing which distinguishes the living system is that complexity of its organisation, and with the increasing interest in nucleic acid chemistry, some of us may believe that if this system is put together it will work. (P. 372)

The attempt to decompose the cell and study the reactions in isolation did not directly reveal the mechanisms critical to fermentation. Moreover, the research led to systematic errors in the understanding of fermentation, including the exclusion of phosphates from the main fermentation process. These errors were overcome only by researchers whose focus was more directed at the physiological needs of the cell, and evidence about how whole cells behave. Thus Embden saw that lactic acid fermentation in muscle juice did not use straight glucose, but required an intermediate form. He also saw that inorganic phosphate was produced in fermentation, and he inferred that the intermediaries probably were phosphorylated. Meyerhof developed models of the physiological processes in which the chemical mechanisms underlying the process could not yet be specified.[16] From the perspective of developing reasonably simple models, phosphorylated intermediaries seemed extraneous and unnecessary. They certainly compromised linearity. In the end researchers discovered that the system responsible for fermentation was *not* nearly decomposable: interaction and organization were critical to the process.

This research program is thus quite different from what we saw in Chapter 6. Instead of finding a component, or a center, exercising an autonomous function, and then proceeding to a lower level to explain its operation, researchers here stayed at essentially the same level and sought to identify the array of components involved in fermentation and to understand their organization. This is what we call *first order interaction*. Instead of isolating independent "organs" whose behavior is relatively independent, first-order effects require an understanding of organization and interaction. This is another alternative, in addition to first order independence, represented in Figure 7.6. The central tasks involve isolating components, their interactions, and the organization that makes those interactions possible. Thus, in explaining fermentation, we eventually must turn to the specific reactions, the enzymes that facilitate them, and the interactions between various stages in the process. This is essentially a functional analysis of the system.

What is critical in understanding an integrated system is the willingness to go beyond what can be learned by studying component reactions in isolation. One must attempt to develop comprehensive models that integrate different processes in accounting for systemic behavior. Such synthetic approaches are obviously fallible. Gustaf Embden, for example, was considered by some of his contemporaries to be excessively speculative. An analytical method, though, is also fallible. Indeed, as we have seen in assuming decomposability or near decomposability, the analytic method may lead us to miss critical interactions. There is no guarantee that the processes are as independent as the analytic method assumes, and as a result it is possible to overlook important features of the system by insisting on an analytical approach. Integrated systems may not escape our observations and understanding altogether, but they may require that we finally accept a failure of decomposition and localization.

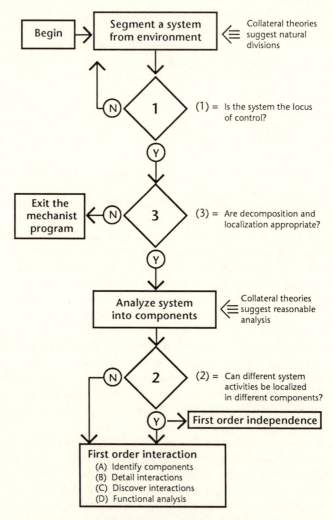

Figure 7.6. A Second Outcome of Decomposition and Localization. Attempts at decomposition and localization can also result in the recognition of a variety of components. In these cases of first order interaction, the organization between components with discrete functions is critical in understanding systemic behavior.

CHAPTER 8

Reconstituting the Phenomena

1. Introduction: Biochemical Genetics

In the last two chapters we have illustrated the use of localization and decomposition in developing models of complex systems. In all the cases we have discussed, the development of explanatory models was significantly constrained by lower-level theories as well as systemic behavior. These constraints varied in relative strength. In Chapter 6 the initial models of brain function incorporated simple decomposability. Linguistic functions on the one hand, and memory functions on the other, were assumed to be functionally independent and discretely localized. However, research did not limit itself to simple localization. The initial localizationist models were followed by an attempt to reveal underlying physical mechanisms. Decomposability nonetheless was retained, as providing at least a first-order approximation of systemic function. Both higher-level, behavioral, constraints and lower-level, physical, constraints were significant in the development of the resulting models of system behavior and organization. In these cases the lower-level constraints on the resulting model were largely correlational and empirical. Though important, they were relatively weak. The primary constraints on the structure of the resulting model were imposed by the higher level. Cognitive models of memory function, or of linguistic processing, were projected top-down onto the physical level.

In Chapter 7 we turned to research into fermentation. There was once more an initial commitment to localization and decomposition, but it eventually resulted in a model emphasizing integration and organization of physical components rather than decomposition. Here again there was initially a commitment to discrete physical units realizing distinct higher-level functions. Single enzymes were supposed to mediate complex biochemical activities; however, the organization revealed by research into fermentation eventually showed that the biochemical mechanisms provided an interactive and integrated system, rather than one that was simply or nearly decomposable. Once again there were higher-level constraints on the models. Fermentation, including the formation of lactic acid and alcohol, was the overall process to be explained, and it was important that a model embody biologically realistic processes. Moreover, there were important empirical constraints. We emphasized, particularly, the requirement that intermediaries be isolable and that their rate of fermentation be commensurate with the rate of the overall process. The lower-level constraints in Chapter 7 are far more restrictive than in the cases discussed in Chapter 6. They imposed strict limitations on the actual structure of the resulting explanatory models, by limiting biochemical processes to simple chemical reactions or sequences of chemical reactions. As a result of the attempt to satisfy the constraints simultaneously, linearity was abandoned, and with it, even near decomposability.

The case that will be at the focus of the current chapter, biochemical genetics, has substantial parallels to the cases in both of the previous chapters. Two traditions converged in influencing the development of biochemical genetics; they provide, respectively, the higher- and lower-level constraints on the development of the field. The first tradition consisted of classical Mendelian investigations into the structure and organization of the genetic material. With the "rediscovery" of Mendel's experimental results at the beginning of the twentieth century,[1] there was a commitment to a particulate model of inheritance, with segregation, dominance, and independent assortment. There was also a commitment to "autonomy" in the expression of genes. Jointly, these Mendelian principles were tantamount to simple decomposability. Deviations from these principles were noticed almost immediately, and they provided the basis for a rich characterization of the phenomena and mechanisms of inheritance in the hands of the Morgan school. As the research program developed, it became clear that simple decomposability could not be maintained. As we will see in the following section, with this recognition it also became apparent that some of the central phenomena revealed in the Mendelian program could not be explained in Mendelian terms, but required a lower-level explanation.

The second tradition influencing biochemical genetics was research into biochemical pathways and their significance for development. While this research included work in physiological chemistry, it focused on work connecting genetic differences and development. This method incorporated assumptions paralleling those affecting research into fermentation. In particular, it required that the basic processes involved in developmental models be known chemical reactions; that intermediaries in metabolic processes be independently isolable; and, initially at least, that there be linear organization. As a result this method imposed strong physical constraints on the resulting models.

The synthesis of these two traditions in the development of biochemical genetics placed strong independent constraints on the resulting genetic models from both levels; models of gene action had to reflect realistic biochemical processes and explain the phenomena uncovered by the Morgan school. As in the case of research into fermentation, satisfying these constraints eventually led to models that no longer embraced linearity, or, therefore, decomposability. We were faced again with systems that are exceedingly complex in terms of organization. But in this case the final result was a reconceptualization of the problem, which led to a new understanding of the phenomena to be explained—a conception requiring that the phenotype be understood in biochemical terms. This *reconstitution of the phenomena* reinstituted decomposability and localization.

2. CLASSICAL GENETICS

In their simplest textbook form, the crucial experimental results attributed to Mendel, and supposedly rediscovered in the early twentieth century, are segregation and independent assortment: alternative alleles maintain their independence through sexual crossings, and genes are distributed to sexual gametes in a random mixture.[2] If true these principles would insure that genes are discrete genetic units. Neither of these results—commonly termed "Mendel's Laws"—commits us to the functional, or physiological, independence of the genes. In particular, neither demands that there be single genes sufficient for the production of unique traits, or that genes affect only single traits. That is, genetic independence as observed in transmission does not guarantee that the relation between genes and observable phenotypes will be simple or straightforward. As far as these principles are concerned, there may be many genes affecting each trait, and their influence may be complex and context sensitive (cf. Morgan et al. 1915, ch. 9).

Historically, though, the experimental paradigm *was* taken by advocates of Mendelism to indicate that, as a general rule, characters developed autonomously. It is not possible to understand classical genetics without understanding the role this assumption played in its experimental paradigm, the breeding experiments with *Drosophila melanogaster*. The assumption is nicely illustrated in Morgan and Bridges's (1919) extended examination of gynandromorphs—mosaic *Drosophila* with some structures that are characteristically male and some characteristically female. Commonly these hybrid gynandromorphs will be male on one side of the body and female on the other and will retain, bilaterally, the sex-linked characteristics of their parents. Figure 8.1, redrawn from Morgan and Bridges (ibid.), depicts hybrid individuals that are bilaterally distinguished. In one case the fly is male on the right; in the other, on the left (for details consult the accompanying text). Morgan and Bridges observe:

> A striking fact in regard to these gynandromorphs is that the male and female parts and their sex-linked characters are strictly self-determining, each developing according to its own constitution. No matter how large or how small a region

may be, it is not interfered with by the aspirations of its neighbors. (Ibid., p. 112; cf. Sturtevant 1965, p. 101)

The conclusion is that development is controlled by genes that are relatively impervious to influences from other cells in the same organism, no

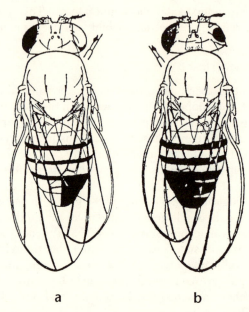

a   b

Figure 8.1. Two Gynandromorphs of *Drosophila*. In these cases of bilateral differentiation, one gynandromorph (a) exhibits male characteristics on the right and female on the left, while the other (b) exhibits male characteristics on the left and female on the right. Male characteristics result from a haploid constitution; female, from a diploid constitution. The right side of gynandromorph (a) is male (note the coloring of the abdomen), with a characteristically smaller size and shorter wings. The white eye results from an elimination of one X chromosome, allowing white to be expressed. The left eye is wild type, resulting from the combination of two chromosomes, coding for white and sable. The right side of gynandromorph (b) is female, with longer wings and a larger size (again, notice the color of the abdomen). The left eye is vermilion and of normal size, resulting from a single chromosome. The right side carries the dominant genes for wild-type (red) eye color and a bar configuration. (Redrawn from Morgan and Bridges 1919, p. 36.)

matter how different they may be in their internal constitution. Sewall Wright later expresses the view in genetical form:

> There is considerable correlation between the gene as a block within which crossing over does not occur and the gene as an apparent physiological unit. Multiple alleles in general affect the same characters and frequently seem to differ only in the degree of effect. . . . On the other hand, there is little or no tendency for genes that are close together in the linkage system to be similar in effect. . . . In general the effect of replacing one allele by another is as independent of replacement in neighboring loci as in any other parts of the genetic system. The translocation of a gene, whether type or mutant, to another region has no apparent effect on its physiological activity. (1941, pp. 490–91)

Even within a single cell there is minimal interaction between genes. Wright concludes that the gene is an "unbreakable physical and physiological unit" (ibid. p. 491). This is direct localization: The genetic system underlying the expression of biological traits is simply decomposable, and each trait is under distinct genetic control. These distinct characters are, in turn, localized in different regions of the chromosomes.

Neither independent assortment nor autonomy were to stand uncompromised for long after their rediscovery. It was only a matter of months until the purity of Mendel's laws was tarnished by laboratory results. In fact, it would be nothing of an exaggeration to say that deviations from the abstract principles were most important for the development of Mendelian genetics, and that these deviations naturally led to problems that required a synthesis of Mendelian and developmental genetics. This synthesis would be realized fully only in the middle decades of this century.

Carl Correns (1864–1933), one of the "rediscoverers" of Mendel's laws, took note of both incomplete dominance and linkage in the very year he published his first paper on Mendelism. In hybridization experiments with *Hyoscyamus*, Correns found that the first-generation hybrids showed an intermediate color rather than that of either parent, a discovery that challenged dominance:

> The rule [of dominance] can only be applied to a certain number of cases, for the present only to those in which one member of the character pair dominates, and for the most part probably only to racial hybrids. That all pairs [in] all hybrids follow it is quite out of the question. (1900, p. 167; quoted in Olby 1985)

Correns (1903) concluded that the explanation of segregating characters and the phenomenon of dominance should finally be explained in terms of chemical processes. Thirty years later Sewall Wright made the point most clearly: "It is clear that dominance has to do with the physiology of the organism and has nothing to do with the mechanism of transmission" (1934, p. 24; cf. Bateson 1909 and Olby 1971).

Deviations from independent assortment were at least as important and are particularly salient in their influence on T. H. Morgan (1866–1945). Morgan was initially skeptical of Mendelism, as well as of the attempt to synthesize it with the chromosome theory (cf. Allen 1978, pp. 125ff.). The identification of genes with discrete segments of chromosomes required that linkage, or "coupling," of traits be common, though it did not appear to be the case.[3] Moreover, Morgan's training as an embryologist left him ill-disposed toward preformed characters in the germplasm—a view he regarded as a revival of preformationism. In 1910, though, Morgan observed a white-eyed mutant male among his *Drosophila*. After cross-breeding the white-eyed male with a normal, red-eyed female, the first generation was uniformly red-eyed. Matings between siblings produced the three to one ratio that would be predicted on Mendelian grounds, assuming that the white-eye is a simple recessive. This confirmation of segregation, however, was not what most struck Morgan. He noticed that *all* the white-eyed offspring were male, though the offspring were equally divided between male and female. The overall ratio was two red-eyed females to one red-eyed male to one white-eyed male. Assuming a chromosomal basis for sex determination, this "sex-limited" inheritance could be explained if the Mendelian factors for eye color were localized on the chromosomes determining sex. As Morgan found more mutants, and two more with sex-limited inheritance, his skepticism of the chromosome theory vanished.

Under the leadership of Morgan, C. B. Bridges, H. J. Muller, and A. H. Sturtevant, the *Drosophila* labs confirmed that genes do not assort independently and are not irresolvably linked. There is significant linkage between traits, but it is incomplete; though traits are separable, as shown by the fact that in a small proportion of offspring they are not transmitted together, that proportion is far less than a simple aggregative model satisfying independent assortment would require. The real work of the Morgan group then began. The researchers set about systematically to map the distances between genes on the assumption that variations from independent assortment indicated the linear distance between linked genes on the chromosome: the greater the distance of two genes on a chromosomal map, the more frequent the recombination between them. An additive model displaced an aggregative one.

During these early years, as we have said, there were investigators who held out for a simple model with minimal interaction between genes; that is, for autonomy. Two lines of research showed that the assumption of autonomy could not be sustained; both came from A. H. Sturtevant (1891–1970), then working in Morgan's lab. The first arose from work with gynandromorphs and provided a demonstration of interaction between distinct cells in development. The second, commonly referred to as *posi-*

tion effect, was a clear demonstration of interaction between genes within a single cell, based on proximity; that is, it showed that there were interaction effects depending on simple spatial contiguity on the chromosome. As a result of these findings, the hopes for a simple, autonomous model of gene action were crushed.

Sturtevant (1920) showed that eye color in *Drosophila* could be altered when under the influence of other tissues. Working with a gynandromorph whose head showed paternal characters and whose inherent genetic constitution would require that it have vermilion (v) eyes, Sturtevant observed that it in fact developed a wild-type color (v +). The color of the eyes deviated systematically from what would be predicted on the basis of the gynandromorph's intrinsic constitution, and it did so in the direction of the maternal constitution. Since he knew eye color was a sex-linked trait, Sturtevant concluded that the nonvermilion color was not autonomously determined by the genetic makeup of the tissues constituting the eye, but was under the influence of the wild-type body tissue. He insightfully speculated that this was due to some *diffusable substances*. For developmental purposes it was clear that autonomy was compromised.

Sturtevant also uncovered significant interaction between genes within the cell. Working with a variant form of *Drosophila* (Bar) which exhibited a reduction in the number of facets in the eye, Sturtevant found yet another variant (Double Bar) which suffered an even more extreme reduction in the number of facets (see Figure 8.2.). Sturtevant and Morgan confirmed that this extreme variant was the result of unequal crossing-over; that is, recombination brought genes normally on *homologous* chromosomes into adjacent positions on a *single* chromosome, and the contiguity of the genes caused the accentuated reduction in the eye. Sturtevant (1925) saw that this position effect could not be accommodated without interaction of a nonadditive nature between genes. It was not only the relative dose of a gene that mattered, but their relative placement on the chromosome. Sturtevant and Beadle (1939) were later to suggest that the correct explanation of this too would be found at the chemical level. Garland Allen comments on the importance of Sturtevant's discovery:

> The idea of position effect was a notable departure from the pure Mendelian view that genes are independent units and thus not influenced by other genes associated with them. . . . The concept of position effect, like that of modifier genes, introduced into the hereditary process a greater degree of complexity than the older notion, derived from some of the early Mendelians, that saw genes as rigidly determined "characterlets." Now, even *position* was seen as affecting the phenotypic expression of a gene. (1978, p. 241)

Allen is certainly right that position effect is a notable departure from the "pure Mendelian view," though the acknowledgment of diffusable sub-

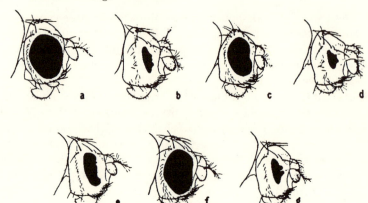

Figure 8.2. Position Effect in the Eye Structure of *Drosophila melanogaster*. The relative area reflects differences in the number of eye facets, seen from the side: (a) Normal, wild type female, (b) Homozygous bar female, (c) Heterozygous bar female, (d) Double bar male, (e) Homozygous infrabar female, (f) Heterozygous infrabar female, (g) Double infrabar male. The bar and double bar forms result from unequal crossing over and are not true mutations. (Redrawn from Sturtevant and Beadle 1939)

stances had already compromised any hope of autonomy in phenotypic expression. Even at this point, and in the absence of a satisfying explanation of the position effect or the nature of diffusable substances, it was clear that interaction between genes discredited autonomy. There simply was no properly *genetic* explanation of this interaction.

The recognition of interaction between genes thus came in a variety of forms. There were many genes affecting each trait, there was important interaction between genes, and there were epigenetic mechanisms at work in controlling development. While speaking before the Royal Society in 1922, Morgan acknowledged the point as clearly as anyone could.

> These ultimate units [genes] are not necessarily to be thought of simply as the representatives of each part of the organism, for every part of the organism must result from the activity of a large number of elementary units. . . .
>
> The evidence has given us a glimpse at least of processes that are so orderly and so simple as to suggest that they are not far removed from physical changes and the order of magnitude of the materials is so small as to suggest that its component parts may come within the range of molecular phenomena. If so, we may be well on the road to the promised land where biological results may be treated as physical and chemical events. (cited in Carlson, 1966, p. 85)

The breeding program of the Morgan school was successful in unraveling the hereditary mechanisms, but it proved relatively ineffective in revealing the developmental mechanisms. There was no explanation of the position effect, and geneticists were at even more of a loss in dealing with Sturtevant's diffusable substances. The next task for genetics was to investigate the structure of genetic control. This meant treating it as a matter of chemical action.

3. Developmental Genetics

Even as the Morgan group was revealing the mechanisms of inheritance, and forging the basis for a Mendelian triumph, there were figures such as Lucien Cuénot (1866–1951), A. E. Garrod (1857–1936), J.B.S. Haldane (1892–1964), and Sewall Wright (1897–1990) who were concentrating on the role that the hereditary materials played in controlling biochemical reactions in the cell. Though research into the biochemical basis of genetic control fell into relative neglect from roughly 1910 through 1935, it was subsequently revived in the landmark work of George Beadle and Boris Ephrussi (1936, 1937).[4] This work revealed a complexity in the biochemical control of development that had remained obscured within the program of classical genetics.

The work of Garrod and Cuénot, as well as much that preceded and followed it, employed assumptions paralleling those generally assumed in biochemistry and discussed in the previous chapter. In particular it was assumed that the basic reactions regulated (whether directly or indirectly) by genes should correspond to known chemical reactions of relatively simple sorts. Ideally these would take the form of enzyme-mediated reactions. This was a natural assumption for Mendelian genetics. For example, differences in pigmentation naturally led to explanations in terms of the presence or absence of biosynthetic enzymes. As Robert Olby said, "An association between the facts of Mendelism and those of biochemical individuality and metabolism was seen by the early Mendelians as natural" (1974, p. 124). It was further assumed that though the reactions controlling development would be complex, the isolation of intermediaries would provide evidence concerning the specific sequence of reactions. They would therefore provide indirect evidence concerning the genes that exercised control over these reactions.

These assumptions and their implications for Mendelian research can be illustrated in two cases. The first concerns work by Archibald Garrod in the first decade of the twentieth century on the genetic basis of metabolic deficiencies. Garrod's work is especially useful in revealing the importance of these two constraints on biochemical genetics during the period and the significance of biochemical constraints and methodology. The second case focuses on work conducted by Boris Ephrussi and George

Beadle in the 1930s, building on the work of Sturtevant. This research began to unravel the complexity of genetic control and ultimately influenced the seminal work of Beadle and Tatum that we will discuss in section 4.

Archibald Garrod and Alkaptonuria

The study of alkaptonuria, a harmless condition marked by a blackening of urine on exposure to air, provides us with a useful starting point.[5] As Olby (1974) points out, it was not altogether clear in the period prior to Garrod's work just how useful the study of alkaptonuria would be in investigating normal metabolic pathways. This is because the most plausible chemical changes in deriving alkapton—or homogentisic acid, the darkening substance—involves changes that did not meet the assumption that metabolic reactions must be limited to known chemical transformations. The chemical structure of alkapton was discovered in 1891 by M. Wolkow and E. Baumann. Wolkow and Baumann noticed that the excretion of alkapton was significantly increased if tyrosine was fed to patients, and they concluded that alkapton must be a result of the oxidation of tyrosine. As depicted in figure 8.3, the reaction required a migration of the hydroxyl group as well as an oxidation. Because the former was a reaction not known at the time, Wolkow and Baumann concluded that it was not a simple chemical reaction and therefore attributed the alteration to the activity of microorganisms in the gut. Wolkow and Baumann remained remarkably unconcerned over *how* the microorganisms accomplished the transformation of tyrosine into alkapton; the important conclusion was that the transformation was *not* a consequence of standard metabolic action, but an anomalous effect due to the influence of invading organisms.

By the time Garrod undertook his studies, biochemists had come to accept side-chain migrations of the sort the model of Wolkow and Baumann required, and so they were acceptable as chemical reactions. Garrod confirmed Wolkow and Baumann's conclusion that homogentisic acid was the cause of alkaptonuria. He had been working with a child who had exhibited the condition since 1898, and by the spring of 1901 a second child who also had alkaptonuria was born to the family. After some questioning Garrod found that the parents were first cousins. This set the stage for a Mendelian interpretation of the disease. Bateson and Saunders (1902) pointed out that such marriages would be most likely to reveal rare and recessive characters. Garrod happily embraced the explanation:

> The mode of incidence of alkaptonuria finds a ready explanation if the anomaly in question be regarded as a rare recessive character in the Mendelian sense. Mendel's law asserts that as regards two mutually exclusive characters, one of

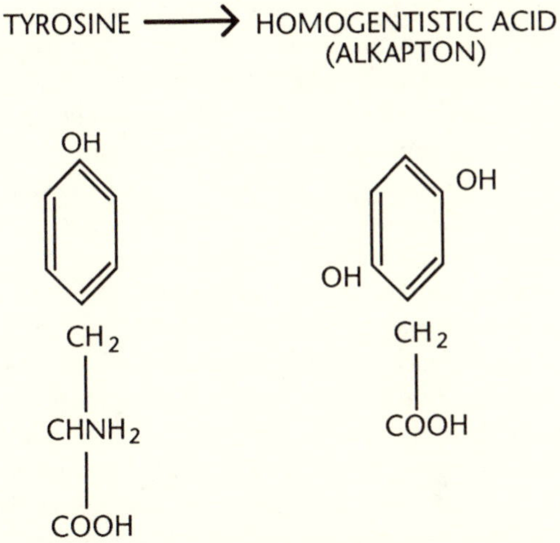

TYROSINE $\longrightarrow$ HOMOGENTISTIC ACID
(ALKAPTON)

Figure 8.3. A Structural Model for the Derivation of Alkapton from Tyrosine. Wolkow and Baumann (1891) proposed that alkapton (Homogentisic Acid) was derived from tyrosine. This required the migration of a hydroxyl group. Because this was, at the time, a structural change that was unacceptable from a chemical standpoint, it was therefore ruled out.

which tends to be dominant and the other recessive, cross-bred individuals will tend to manifest the dominant character, but when they interbreed the offspring of the hybrids will exhibit one or other of the original characters. (1908, p. 5)

Garrod acknowledged that the actual proportions found in studies of families with alkaptonurics was not what would be required by Mendelian principles, but he maintained nonetheless that it was a Mendelian recessive, primarily because "albinism, which so closely resembles it in its mode of incidence in man, behaves as a recessive character in the experimental breeding of animals" (ibid., p. 6).

What is important for our purposes is not just that Garrod thought alkaptonuria was a recessive character, but the way he integrated this with a biochemical picture of metabolism. Appealing explicitly to Bernard, Garrod said,

The view is daily gaining ground that each successive step in the building up and breaking down, not merely of proteins, carbohydrates, and fats in general, but even of individual fractions of proteins and of individual sugars, is the work of special enzymes set apart for each particular purpose. (Ibid., pp. 1–2).

Chemical intermediaries in metabolism would be likely to have a transient existence, being broken down almost immediately; but if, for some reason, catabolic or metabolic processes are interrupted, the intermediates should be excreted. Abnormal excretions should therefore provide direct evidence concerning metabolic intermediaries, and indirect evidence concerning the metabolic processes themselves. If it is assumed that the presence or absence of an enzyme is controlled by Mendelian factors, then there is a natural explanation for the "inborn errors of metabolism" such as alkaptonuria in terms of specific genes controlling enzymatic reactions. That is precisely what Garrod proposed.

Garrod's proposal clearly embodies both assumptions noted above. To begin with, it assumes that the explanation of alkaptonuria must be carried out at a chemical level, appealing only to reactions of a known character. Enzyme action is assumed to be specific and limited. Moreover, once an enzyme system is blocked, Garrod assumes that the reaction it mediates is unlikely to be accomplished by an alternative mechanism: "If the conception of metabolism in compartments, under the influence of enzymes, be a correct one, it is unlikely, *a priori*, that alternative paths are provided which may be followed when for any reason the normal paths are blocked" (1908, p. 2). Secondly, once an enzyme is blocked, Garrod assumes metabolic intermediaries will be detected in excretions. Alkaptonuria must be seen "as an arrest rather than as a perversion of metabolism" (ibid., p. 217). It is essential to Garrod's view of the metabolic pathway that neither tyrosine nor homogentisic acid be normally excreted and so have to be broken down in the more common cases. He conjectures, therefore, that the benzene ring is split by an enzyme that is simply absent in the congenital alkaptonuric. Genes regulate metabolism by controlling these enzymatic reactions; when the dominant gene is absent, so is the enzyme.[6] Lack of the enzyme blocks the normal process, resulting in intermediaries which are ultimately excreted. This is a natural experiment in which the metabolic processes are interrupted, yet it says very little about the complexity of metabolism and catabolism. It is in fact consistent with an assumption that the process is linear, degrading more complex proteins in a stepwise fashion. Garrod seems to have entertained nothing else.

Garrod's investigation of the inborn errors of metabolism preceded those of the Morgan school and fit largely within a more medical context with a different research agenda. According to the program of the Morgan school, phenotypic traits were explained in terms of localized genes, and the problems surrounding gene expression were relegated to a secondary status. The focus was on what the statistical distribution of traits within breeding populations could reveal about chromosomal structure and the localization of genes. Garrod's work, by contrast, focused on how genes were expressed; the inborn errors of metabolism were of interest precisely because they provided information about the mechanics of gene expression. While Garrod certainly assumed that the responsibility for both normal and pathological development lay in the genes, proximate control was attributed to enzymes. Biochemical pathways were thus interposed between phenotypic traits and genes, and the presence or absence of an enzyme was what received explanation in Mendelian terms. The details of chromosomal structure were of secondary importance.[7]

Genetic Control of Development

A similar emphasis on gene expression and development, with parallel assumptions, can be seen in subsequent work on genetic control carried out in the mid-1930s by Boris Ephrussi and George Beadle.[8] The nonautonomous control of eye color in *Drosophila* discovered by Sturtevant promised a direct route to understanding gene action and the underlying means of control (for useful reviews, see Beadle 1945, pp. 33 ff.; Ephrussi 1942). It was known from the crossbreeding of mutant forms that the red eye of normal, or wild-type, *Drosophila* was the result of two components: a red pigment and a brown pigment. Either could be inhibited by mutations. The formation of brown pigment could be blocked (to varying degrees) by mutations at at least four loci, leaving the red pigment and a bright red eye. The red pigment also could be absent, and its formation inhibited by mutations, leaving only brown pigment and a brown eye. Finally, both pigments could be blocked, leaving no pigment and a white eye. The recognition of nonautonomous control for eye color inspired transplant studies by Beadle and Ephrussi (1936, 1937) in which the imaginal disks from the eyes of mutant larvae of *Drosophila melanogaster* were implanted into the abdomens of normal larvae, and vice versa. The resulting eye color of the implanted disk would be influenced both by the intrinsic genetic constitution of the eye disk and by the surrounding tissue. Their work showed that the color deviated systematically from what would be predicted on the basis of the genetic composition of the larvae alone.

As Table 8.1 indicates, v imaginal disks transplanted into v+ larvae developed a color appropriate to v+, and v+ disks implanted into v larvae maintained a v+ color. Parallel results were also seen with a second mutant color, cinnabar (cn). These results were consistent with the view that v and cn were simple recessives, and the wild-type v+ a Mendelian dominant trait. Normal development followed if either the imaginal disk or the host tissue contained v+ genes. The most significant cases were the reciprocal transplants of v and cn. A v disk implanted on a cn larvae developed v+ color. Conversely, however, a cn disk implanted on a v larvae developed a cn color. These results were incompatible with an interpreta-

Implanted Eye Disk	Host Type	Color of Implant
Vermilion (v)	Wild Type (v+)	Wild Type (v+)
Vermilion (v)	Cinnabar (cn)	Wild Type (v+)
Cinnabar (cn)	Vermilion (v)	Cinnabar (cn)
Cinnabar (cn)	Wild Type (v+)	Wild Type (v+)
Wild Type (v+)	Cinnabar (cn)	Wild Type (v+)
Wild Type (v+)	Vermilion (v)	Wild Type (v+)

Table 8.1. Summary of Transplantation Experiments by Beadle and Ephrussi (1936, 1937). It is particularly important that an implanted cn disk develops a cn phenotype under the influence of a v body type, but that an implanted v developed into a v+ phenotype under the influence of a cn body type. This implies that the relevant genes control sequential reactions in development, rather than being allelic variants, or providing independent contributions.

tion treating v and cn as allelic variants or as providing independent contributions. They could not be accommodated without positing a sequential, or linear, organization. Beadle and Ephrussi concluded that there were at least two intermediaries implicated in the formation of pigment in a normal eye and that they occupied sequential positions in the metabolic chain. Ephrussi wrote in 1942:

> There are two different substances, one responsible for the change from vermilion to wild type and the other for the change from cinnabar to wild type. The wild type lymph contains both these substances. The lymph of the mutant cinnabar contains only one of them, namely the substance responsible for the change from vermilion to wild type. The mutant vermilion contains none of these substances. . . . [The] two substances are formed in the course of a single chain of reactions, of which the v+ substance represents the first and the cn+ the second link: $\Rightarrow$ v+ $\Rightarrow$ cn+. (pp. 329–30)

There were at least two diffusable substances, rather than one, and they were organized sequentially. The v mutant failed to carry out the first reaction and therefore could not synthesize the second of the diffusable substances. As a host the v mutant could not supplement the cn implant, which also could not carry out the second transformation. The cn mutant could perform the first reaction, but not the second. As a host it therefore could carry out the first step in the reaction and could "repair" the eye color of the v implant. In 1939 Beadle and Ephrussi concluded that the blockage in reactions was due to a lack of specific enzymes (cf. Olby 1974, p. 141).

Beadle later remarked that the explanatory scheme involved in his work with Ephrussi relied on the assumptions that there were at least two gene-regulated chemical reactions involved and that the intermediaries would accumulate when subsequent reactions were blocked. Whether or not this was a reliable interpretation of Beadle and Ephrussi's earlier assumptions, their work did reveal a structure in the synthesis of pigments and indicated that genetic action needed to be conceptualized at a more discriminating level. Eventually Beadle and Tatum (1941a) came to portray the synthesis of brown eye pigment as in Figure 8.4.

As with Garrod's work, Beadle and Ephrussi's model manifests the three assumptions involved in biochemical research. It incorporates only chemical reactions mediated by specific enzymes; genes are conceived as acting to control specific reactions, either because they are enzymes or because they control enzymes. It assumes that disruption of these meta-

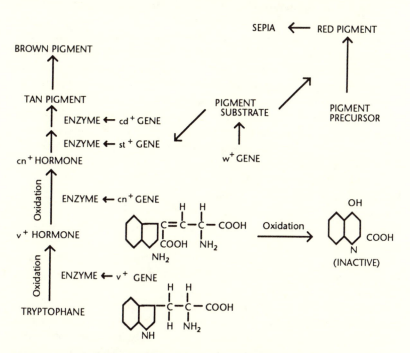

Figure 8.4. The Synthesis of Eye Pigment in *Drosophila*. Notice, in particular, that the transformations explicitly represented in the structure diagrams are simple oxidations and that the overall organization is linear. As in Beadle and Tatum's earlier representation of the process, inactivation of the v + or cn + genes would block the formation of brown pigment, leaving red pigment in the eye. (Beadle and Tatum 1914a, p. 114.)

bolic pathways will result in a residual substance detectable in mutant eye colors. The residual pigment in **v +** or **cn +** mutants is thus seen as the result of blocking a pathway in which the usual result would be the brown pigment contributing to the normal red eyes. The organization the model uses is linear, and because of the linear organization the disruption of a gene will affect only what is subsequent to it in the synthesis of the pigment. Again, this is near decomposability. As with Garrod's work, the decomposition depends on an understanding of the biochemical processes and not simply on distributions of phenotypic traits.

4. One Gene/One Enzyme

In light of the work from the Morgan school and developmental genetics, the general strategy of Beadle and Tatum's classic experimental work on *Neurospora* is reasonably straightforward. However, despite its being a natural extension of a classical paradigm, it was to have revolutionary consequences. Beadle and Tatum recognized the limitations of the classical paradigm; in particular, they saw that the methodology almost inevitably left them unable to see anything but genes affecting superficial characters. This is because the methodology of Mendelian genetics committed it to beginning with the character and working backward to determine the sequence of reactions. The phenotype and statistical patterns of inheritance were the only acknowledged constraints. In practice this imposed several limitations. Breeding experiments could reveal genetic factors only when there were alternative alleles, and then only if those alleles had detectable phenotypic effects. This meant that alleles with small effects would be likely to escape detection, and, correspondingly, that those with the largest effects would likely be lethal. Since terminal reactions were those most prone to modification by nonlethal mutants, this also meant that the classical paradigm would naturally be led to focus on genes controlling these terminal reactions. The underlying complexity of developmental processes, and their genetic control, was going to elude the grasp of classical work.

The conjunction of Mendelian genetics with research on biochemical genetics restructured the problem into one also constrained by the nature of the biochemical processes underlying phenotypic expression. This required an experimental methodology and a model system that could avoid the problems facing Mendelian genetics. For this Beadle and Tatum drew on the earlier work with Ephrussi. They set about to investigate the nutritional requirements of *Neurospora*, recognizing that genes must control the biosynthetic processes in nutrition and, further, that these processes would exhibit considerable complexity at the genetic level. Their goal was to uncover the underlying genetic organization by investigating different nutritional requirements of mutant strains. What finally emerged from the research was a reconceptualization of the phenomena needing to be explained. This is what we call *reconstituting the phenomena.*

Beadle and Tatum retained the constraints characteristic of work in biochemistry and of the earlier work in the developmental genetics of *Drosophila* (cf. Tatum 1959, p. 1712). They also maintained the assumption that genes control biochemical processes through specific enzymes:

> From the standpoint of physiological genetics, the development and functioning of an organism consist essentially of an integrated system of chemical reactions controlled in some manner by genes. It is entirely tenable to suppose that these genes which are themselves a part of the system, control or regulate specific reactions in the system either by acting directly as enzymes or by determining the specificities of enzymes." (1941b, p. 499)

Genes are specialized in their action, and their heterocatalytic products will be equally specialized. As Beadle put it four years later:

> Each nucleus of those organisms sufficiently advanced in the evolutionary scale to have nuclei contains many thousands of genes. . . . Each of these thousands of gene types has, in general a unique specificity. This means that a given enzyme will usually have its final specificity set by one and only one gene. (1945, p. 19)

Specificity of enzymatic activity is methodologically important for reasons we have already seen in conjunction with Garrod's work. Specificity suggests that there will be no alternative pathways to mediate effects once a gene is inactivated. This makes it possible, in principle at least, to detect the consequences of such inactivation. It also makes enzymes a natural unit in terms of which to understand genetic action.

Beadle and Tatum also retain the assumptions that biosynthesis must occur via a series of sequential steps and that disruption at any given step should inhibit the overall effect. Mutants affecting metabolic functions should therefore be detectable, provided that the deleterious effects are initially overcome. We have already seen that these assumptions were present in the work on pigmentation, where the damaging effects of the mutant genes is what allowed them to be detected.

As we have noted, Mendelian methods were limited to dealing with naturally occurring, nonlethal variations. Beadle and Tatum's approach was designed to overcome these limitations. With this in mind, *Neurospora* was an ideal organism to use in their research. To begin with, it has simple nutritional requirements: it can be grown on a medium containing a carbon source (sugars, starch, etc.), a nitrogen source, inorganic salts, and biotin (a B vitamin); it can also be grown with more complex supplementation. This means that normally it is capable of synthesizing most of what is necessary for growth: nutritional simplicity betokens complexity in biosynthetic processes. Secondly, *Neurospora* can reproduce either sexually or asexually. Asexual reproduction allows for a sizable culture of genetically identical individuals whose nutritional requirements can be evaluated; sexual reproduction enables researchers to carry out Mendelian crosses on individuals resulting from asexual reproduction. Finally, the organism is haploid. This means that there is no masking of genetic effects by dominance relations; a mutant gene will inevitably be expressed, and a gene inactivated by mutation will have the effect of disabling any pathway of which it is a constituent.

Beadle and Tatum induced mutations using X-rays.[9] The goal of this inhibitory study was to bring about gene mutations, with the hope that those mutations would affect the traits under examination. As Beadle explained, the "inactivation of specific genes is equivalent to the chemical poisoning of specific enzymes, with the important difference that genes are highly specific, whereas enzyme poisons are often discouragingly nonspecific" (1945, p. 61). Allowing meiosis to take place, Beadle and Tatum were able to obtain a host of genetically homogeneous spores. After initially growing them on an amply supplemented medium, until a suitably large population was available, they transferred samples to a medium having the minimum amount of supplements sufficient for growth of nonmutant strains. Any strains that did not grow on the minimal medium had to be mutant forms incapable of synthesizing some substance that the normal strains did synthesize. By transferring these mutant strains to a variety of media with different supplements, it was possible to isolate the substances the mutant strains were unable to synthesize.

The initial studies by Beadle and Tatum (1941b) uncovered three mutant forms. All grew on the complete medium, none on the minimal medium. Each in turn could be grown on some medium supplemented in one way or another. One grew with the addition of vitamin B6, another with vitamin B1, and a third with paraaminobenzoic acid. Beadle and Tatum concluded that all three were single-gene mutants; they subsequently obtained a variety of other strains that could not grow without supplementation by some specific vitamin or amino acid. They concluded that the assumption that genes control enzymatic reactions must be correct: "A single gene may be considered to be concerned with the primary control of a single specific chemical reaction" (Tatum and Beadle, 1942, p. 240). As a consequence, "the gene and enzyme specificities are of the same order" (Beadle and Tatum, 1941b, pp. 499–500).[10]

Although initially the biochemical reactions were assumed to constitute a linear chain, linearity was eventually compromised here just as in the understanding of cell metabolism. This can be simply illustrated in the case of the synthesis of the amino acid arginine. Following the same methodology as before, Beadle and Tatum grew a variety of genetically distinct strains of *Neurospora*. Some required supplementation by arginine; some by either arginine or citrulline; yet others by arginine, citrulline, or ornithine. On the face of it a simple two-step process was all that was needed, as illustrated in Figure 8.5. If the second reaction was blocked, then supplementation by arginine was required; if the first, then either citrulline or arginine sufficed. If the synthesis of ornithine was impeded and the two remaining reactions remained intact, then supplementation by any one of the three was sufficient.

However, the situation was more complex than this would suggest. Arginine was subsequently degraded into ornithine and urea, and the urea further broken down into carbon dioxide and ammonia. As a consequence, the overall reaction was better represented as a cycle (see Figure 8.6).

$$
\begin{array}{ccccc}
 & & \text{CONH}_2 & & \text{NH}_2 \\
 & & & & \text{CNH} \\
\text{NH}_2 & & \text{NH} & & \text{NH} \\
(\text{CH}_2)_3 & => & (\text{CH}_2)_3 & => & (\text{CH}_2)_3 \\
\text{CHNH}_2 & & \text{CHNH}_2 & & \text{CHNH}_2 \\
\text{COOH} & & \text{COOH} & & \text{COOH} \\
\end{array}
$$

Ornithine => Citrulline => Arginine

Figure 8.5. The Derivation of Arginine from Ornithine. Each step involves a relatively simple alteration of structure and preserves linearity.

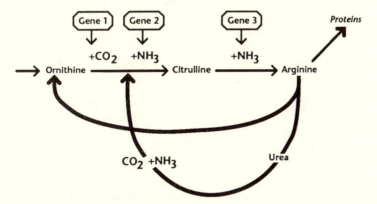

Figure 8.6. A Cyclic Representation. Once again the overall reaction can be better represented as cyclic; the byproduct of one reaction is a substrate in others that occur earlier in the process.

Again, what served as the product of one reaction was a substrate in others that occurred earlier in the process. The normal functioning of the organism, and its classical phenotype, are products of complex and interactive pathways under the control of a constellation of genes. Localization and decomposition appeared once again to have been compromised.

5. CONCLUSION: RECONSTITUTING THE PHENOMENA

Once again interaction and organization were critical. The simple localization of the Morgan school, projected on the basis of phenotypic patterns, did *not* take us to Morgan's "promised land where biological results may be treated as physical and chemical events." It *did*, however, spawn a program of research that gave us an understanding of the inherent complexity of the underlying biochemical processes. Direct localization was eventually foresworn. In part this was a simple consequence of the fact that phenotypic traits were products of many genes in a complex organization. The complexity of gene action was clear even in the *Drosophila* work itself. Though mutations would often have an additive effect, this was not always so. There were modifiers that would have no effect by themselves, but would affect the expression of other mutants. There were other interactions between genes based on location, and even effects between genes in different parts of the body. Moreover, distinct genotypes (such as scarlet, vermilion, and the double recessives) were almost indistinguishable phenotypically. Sewall Wright observed:

> There is usually a simple one to one relation between gene and the substances responsible for immunological specificity. The suggestion is that these substances or at least the active group . . . is a direct product of the gene. . . . In general, however, the relation between gene and such substances as the animal and plant pigments and excretion products is less direct. Many genes, often acting apparently in sequence, are necessary for a particular product. (1941, p. 514)

Autonomy in gene action turned out to be the exception rather than the rule. Decomposition into independent traits failed.

The abandonment of the simple localizationism of the Morgan school was also, in part, the consequence of the integration of the Mendelian program with an independent line of research derived from biochemistry. The breeding program of the Morgan school left a variety of effects, such as dominance and the position effect, which clearly needed explaining but could not be explained in the terms of Mendelian genetics. Some account of these phenomena was needed, and it was natural to look to biochemistry to unravel the mechanisms. The result was an increasingly complex picture of the nature of gene action. Once again, turning to Sewall Wright,

> The reaction chains connecting primary gene action and observed effects on morphological characters must be longer, more ramifying and more heterogeneous than where effects are on intracellular products. Even more indirect are the relations of genes to modes of behavior of the organism as a whole, although there are cases in which there is simple mendelian heredity. (1941, p. 521)

Not only are there generally multiple factors affecting characters, as even Mendelism allowed, but the relations of genes to the characters they produce is "indirect" and "heterogeneous." Gene expression involves interactions and dependencies beyond the reach of Mendelism.

The result of the integration of Mendelian and biochemical genetics was not simply a recognition of a more complex organization where previously a decomposable or a nearly decomposable system had been proposed. Rather, the phenomena of genetics were reconsitituted at the level of biochemistry. Joshua Lederberg explicitly suggested this as the moral:

> Experimental genetics is reaching its full powers in coalescence with biochemistry: In principle, each phenotype should eventually be denoted as an exact sequence of amino acids in protein and the genotype as a corresponding sequence of nucleotides in DNA. (1960, p. 269)

Lederberg's proposal was more radical than might first appear. What traditionally counted as a Mendelian trait, the macroscopically observable phenotypic trait, would be abandoned as the central Mendelian unit, with a shift of the entire analysis to a lower level. The one gene–one enzyme model told us that the level of specificity at which we must understand gene action was at the level of chemistry. In the face of a new vision of the mechanisms, Mendelism's one gene–one *trait* emphasis could be retained by embracing a one gene–one *enzyme* model. We simply had to reconceptualize the relevant traits at a lower level of analysis. Traits too needed to be identified and individuated at the level of enzymes, and the classical phenotype resolved into a complex of traits at that lower level. It was no longer eye colors, but the enzymes that produced them, which become the proper unit for a Mendelian analysis. The observable macroscopic traits that the Morgan school placed at center stage were dissolved under the stronger resolution of biochemical genetics.

Lederberg's suggestion had the merit of simplicity, and in many respects the practice of biology has followed that lead. In some regards the suggestion should appear suspect. We naturally assume that we understand antecedently *what* is to be explained, and an *adequate* explanation is one that conforms to the phenomena. Philosophical predilections to the contrary, once we attend to the development of scientific theories and research programs, it does not appear that our conception of what needs explaining has epistemic priority; instead, it is shaped by the explanations and models we develop.

Reconsitution of the phenomena does not compromise localization and decomposition, or the search for genetic mechanisms, but validates them. It is here that we see the real power of a fully developed mechanistic explanation. A mechanistic approach is not limited to explaining phenomena that are taken as simply "given," but can mandate a revision of the way the phenomena are to be conceptualized and what are given epistemic priority. In the case proposed by Lederberg the reconceptualization of the phenomena accompanies a shift to a lower level (Figure 8.7). Genes are thought of as specific in their action and as acting in relative independence of other genes. There is independence at the level of the enzymes they produce. Conceived in terms of the observable phenotypes, genes have a complex role in development and metabolism, and a complex organization. Simplification followed only on understanding the phenotype differently. A characterization of the phenomena in terms of observable traits was replaced by one couched in terms of biochemical products.[11] This was still localization and decomposition, but this time the reconstitution of the phenomena with a shift to a lower level allowed us to retain localization and decomposition in the face of complex organization.

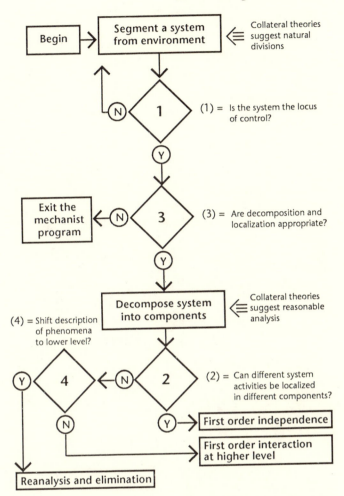

Figure 8.7. A Third Outcome of Decomposition and Localization. An emphasis on organization and interaction can appear to compromise decomposition and localization altogether, as constituent processes are strongly interdependent and organization is critical. However, one possible response, which we call reconstituting the phenomena, leads to a reconceptualization of the phenomena to be explained in terms of the underlying lower-level mechanisms.

PART IV

Emergent Mechanism

The mechanist is intimately convinced that a precise knowledge of the chemical constitution, structure, and properties of the various organelles of a cell will solve biological problems. This will come in a few centuries. For the time being, the biologist has to face such concepts as orienting forces or morphogenetic fields. Owing to the scarcity of chemical data and to the complexity of life, and despite the progress of biochemistry the biologist is still threatened with vertigo.

—A. Lwoff 1950

That the adoption of the mechanistic view has had profound and far-reaching consequences for the whole of society is an historical fact which gives rise to the most divergent opinions. Some commend it as a symptom of the gradual clarification of human thought. . . . Others, though recognizing the outstanding importance it has had for the progress of our theoretical understanding and our practical control of nature, regard it as nothing short of disastrous in its general influence.

—E. J. Dijksterhuis 1950

INTRODUCTION

In Part II we saw that the rejection of localization and decomposition tends to accompany the rejection of a mechanistic program. Providing a mechanism involves describing distinct components, each of which makes a contribution to the performance of the system. This requires both functional and physical independence. In the simplest cases, these components are thought of as making their contributions independently: nature is simply decomposable and embodies an aggregative organization. In slightly more complicated cases, the components are thought to make their contributions sequentially, or linearly, and to retain an integrity of their own: nature is nearly decomposable. As we saw in Part III, a wide variety of organizations may be revealed by beginning with an assumption of near decomposability. The resulting models may not retain the integrity of the components, but may describe what we have termed an integrated system. In such a system nature is at best only minimally decomposable. If organization becomes even more dominant in explaining the behavior of the system, and we appeal less to different and distinctive functions performed by the components, we reach a point where decomposition and localization in any recognizable form have to be surrendered.

While, historically, forgoing decomposability seemed to require giving up mechanistic approaches, this is not the only possibility. Formal modeling techniques have made it possible to explore the behavior of systems in which the components play very minor functions; the explanation of how the system behaves lies in the way these components are organized. Component behavior can be very simple, given a complex and interactive organization. Such *connectionist* systems represent an alternative way of elaborating a mechanist program without assuming decomposability or near decomposability. In Chapter 9 we briefly explore three cases in which researchers have pursued connectionist models and look at the modes of reasoning that support this alternative conception of mechanism.

We begin with a description of the contributions of John Hughlings Jackson, a late-nineteenth-century neurophysiologist who rejected the localizationist programs of Broca and others. Jackson was no vitalist, and dualism played no significant role in his scheme. His opposition to the localizationist program of neo-Phrenologists thus stood in sharp contrast to that of Flourens. Jackson denied discrete modules in the cortex controlling specific functions, just as Flourens did. Jackson did propose an alternative decomposition of the nervous system into lower and higher levels, but, unlike phrenologists or neophrenologists, these levels were not specialized, task-specific modules with discrete functions. Instead, higher levels regulated and orchestrated what was already governed at lower levels. Hughlings Jackson's approach was suggestive as a way of interpreting lesion data, but, given the resources of the time, it was very difficult to develop into a precise theory of how the nervous system worked.

We are now capable of modeling such systems. Until recently researchers in artificial intelligence have taken a computer that processes information in accord with rules to simulate, and perhaps to realize, thought. A new generation of researchers is exploring how connectionist systems might explain a variety of cognitive phenomena. Instead of viewing cognition as involving the processing of information according to rules, cognitive behavior is seen as the "emergent" product of a system that consists of simple units and controlled by simple learning rules. According to this approach, computers are employed not to realize rules or procedures performed on symbols, but rather to solve the equations that characterize the behavior of systems with large numbers of components. These simulations allow the computer to show how behavior might emerge from the interactions of simple components. These connectionist systems are apparently capable of overcoming some of what have seemed to be the greatest liabilities in more traditional artificial intelligence, and they do so *without* assuming that cognitive tasks can be decomposed into discrete subtasks, or that the system is organized into modules; instead, organization and integration supplant compartmentalization.

Our third example of the emergence of connectionist thinking comes from a quite different domain, work on genetic regulation. In recent years biologists have developed increasingly complex models of gene regulation, positing genes having the function of regulating other genes in a localizationist fashion. In an evolutionary framework, however, this becomes quite problematic. The more complex a system, the stronger the evolutionary pressure would have to be to maintain it in the face of mutations. Stuart Kauffman's recent work is directed toward explaining how evolution can maintain such a regulatory system; however, he has ended up *redescribing* the phenomenon to be explained, in something like the manner of the case discussed in Chapter 8. Kauffman develops an abstract, formal model of a genetic system to simulate the regulatory process, a model remarkably like the connectionist systems developed to simulate cognition. These model genetic regulatory networks function by allowing units to excite and inhibit each other until a stable pattern of activation is achieved. Kauffman equates these stable patterns with cell types. Random perturbations will disturb the system's behavior, but it turns out that while it is nearly impossible for selection forces to maintain a large and complex system against mutation in an arbitrarily defined optimal state, some systems will evolve toward a stable state where mutations create only local perturbations and are unable to alter the overall condition of the system. The organization assures a dynamic steady state; moreover, it takes tremendous selection pressure to move the system out of these stable states. This suggests an answer to the original problem, but one that requires changing the question. Rather than asking how selection could maintain a complex regulatory system, Kauffman claims that the regulatory system is inherently stable and does not require selection to sustain it. The result of developing a connectionist system to model gene regulations suggests that what was a pressing issue when the system was assumed to be nearly decomposable does not even require explanation.

In Chapter 10 we place the description of theory development we have built using the cases discussed in Chapters 3–8 within a broader context. We have focused on a number of choice-points confronted in the development and elaboration of mechanistic programs, points which describe the kinematics of theory development. The informal flow chart falls short of a fully dynamic model, though it is a useful step in that direction. It is also useful in suggesting the general directions that a realistic account of discovery might take. Following the lead in Chapter 2, our primary focus has been on psychological constraints and, in particular, on the heuristics of decomposition and localization. In examining the cases in Parts II and III, we have explored how these heuristics came into play in shaping research programs. It is clear from the cases discussed in Part III that the heuristics of decomposition and localization underdetermine the result. There are several directions development takes, depending on other contingencies; the heuristics do not operate in isolation.

In a more speculative vein we will identify three general factors that feature, in conjunction with heuristics, in explaining the dynamics of theory development. These include *phenomenological regularities*, *operational constraints*, and *physical constraints*. We have no detailed account of the character of these constraints, nor of their interaction. A fully elaborated description of how these factors, and others still not identified in any detail, figure in the course of theory development lies beyond the scope of this book. Until these factors are detailed we doubt it will be possible to develop a complete normative account of the history of mechanistic research programs, much less a comprehensive normative account of theory development. Having examined some of the factors affecting theory development, though, we can begin to glimpse what a more adequate account of discovery would look like.

CHAPTER 9

"Emergent" Phenomena in Interconnected Networks

1. Introduction: Dispensing with Modules

The more complex localizationist explanations we examined in Part III are still recognizably mechanistic. Tasks involved in performing a function are divided between components, and system behavior is explained by showing how it is accomplished through the combined performance of the component tasks. Although one might prefer explanations in which the component tasks can be thought of as following a linear, sequential order, so that the contributions of each component can be examined separately, natural systems are not always organized in such a manner. Component tasks are often dependent on one another, so we cannot understand the operation of the system by imposing a linear order on it. Cyclic rather than linear organization occurs when the activity of any given component is dependent on a variety of other components that, in turn, depend on it. In integrated systems, the explanation of the behavior of the whole system depends in a *nonlinear* way on the activities of the components and on the modes of interaction found within the system.

To the extent that organization is important in affecting system behavior, a system is nondecomposable. As we discussed in Chapter 2, there is a continuum of cases. At one end are simply decomposable systems for which the major explanatory task is to identify the components and understand their behavior. These include the kinds of cases considered in Chapter 4, in which a single part is held to be responsible for the behavior of the whole system. Toward the other end of the continuum lie cases of integrated systems in which the behavior of the system is largely due to the interaction of the components. In all of these cases, the operations of the components can, nonetheless, be understood in terms of the operations performed by the whole. Conversely, the behavior of the whole is explained in terms of the behaviors of component parts. There are other systems, yet farther out on the continuum, in which localization and decomposition appear to be hopeless, or even misguided. The hallmark of these cases is that, given a principled structural analysis, the activities of the parts seem to be different in kind from, and so far simpler than, those performed by the whole. The parts can be so simple, in fact, that they do not seem individually to contribute anything of interest to understanding the behavior of the whole; in some cases it is possible to destroy or disable much of the system without significantly affecting performance.[1]

With systems in which the parts do not seem to be performing intelligible subtasks contributing to the overall task, classical mechanistic strategies—and, in particular, decomposition and localization—fall short. What alternatives are there to pursuing a program of mechanistic explanation? In Chapter 5 we considered one possibility: rejecting the mechanistic program and settling for descriptive accounts of behavior. But this is not a strategy for developing an explanation; it is a denial of any explanation. In this chapter we consider an explanatory strategy that abandons localization and decomposition. We leave open whether it constitutes a properly mechanistic approach.

We examine three cases in the following sections. Hughlings Jackson provides a transitional case. Jackson rejects the localization of Broca and proposes a complex, hierarchical model of the nervous system in which control of tasks was distributed over different neural structures at different levels, with higher-level systems regulating and modulating performance of lower-level systems. Our other two cases come from contemporary research, which use newly developed formal mathematical tools for modeling the behavior of complex systems. These two cases are, respectively, models of cognitive performance and genetic regulation. In these cases performance depends primarily upon the interaction of the components in the system. The components do not perform tasks that would appear in a functional decomposition of the system.

2. HIERARCHICAL CONTROL: HUGHLINGS JACKSON'S ANALYSIS OF THE NERVOUS SYSTEM

In the late nineteenth century, John Hughlings Jackson developed a hierarchical model of the nervous system that was intended as a repudiation of the kind of localizationist claims advanced by Bouillaud and Broca.[2] These latter "neophrenologists," as we have seen, constructed localized models on the basis of correlations between neurological lesions and pathological symptoms such as the loss of coherent speech. Losses of specific capacities were traced to injury or destruction of specific regions. While such correlations certainly can be found, they are not as precise as would be required to justify the strong conclusions of these localizationists. The clinical syndromes are not simple, and the neurophysiological disorders rarely correlate precisely with the syndromes. For example, aphasia—in at least one of its myriad forms—is commonly described as a deficit affecting the verbal expression of language, but not affecting cognitive abilities. However, these "expressive" aphasias do not affect the whole range of verbal expression equally and are generally accompanied by a host of other deficits (for example, in reading, writing, and articulation); moreover, some aspects of verbal expression are more affected than others. Likewise, the physiological deficits are not neatly circumscribed. Considerable inference and conjecture are required to draw conclusions concerning the locus of the damage. Expressive aphasias are accompanied, as one would expect, by damage to the third frontal convolution; but damage from either external trauma or stroke is not clearly limited or localized.

One response to this lack of precise correlation between cerebral damage and pathological syndromes is to reject the search for a mechanistic explanation. As we saw in Chapter 5, when Flourens failed to substantiate Gall's correlations between craniological structures and psychological capacities, he denied the localizationist program and gave up the search for mechanisms governing the higher cognitive capacities. This was not an option Jackson could accept. He was committed to understanding the neurophysiological operations of the brain, convinced these would explain the associated psychological deficits. This required making sense of how the brain accomplished its operations through the interaction of its parts. As a good clinician, the complexity of the symptoms was always before him. He was led to a different explanatory approach.

Jackson found the key to developing an alternative interpretation of the operation of the nervous system in a different pattern of deficits, this time in epileptic seizures (see Jackson 1884; Melville 1982). Epileptic seizures, or the more extreme cases at least, are generalized, affecting the entire body to varying degrees. Jackson saw three levels of seizures (cf. Jackson 1884, pp. 57ff.). The first and least severe is analogous to a dream state. The second is accompanied by a loss of consciousness. The third leaves the patient comatose. These three levels increase in severity, with the third most encompassing and the first least so. As a consequence of his complex symptomology, Jackson maintained, the epileptic discharge must begin in a region that affects the body as a whole—including the highest levels, such as volition. It must begin in cortical structures. A mild seizure is the analogue to dreaming. A more severe discharge affects the lower levels and also disrupts consciousness. In yet more severe cases, the functioning of more central structures are also disrupted. In an analogous manner, Jackson thought expressive aphasias deprive the patient of the ability to use complex forms of language, but leave the more automatic uses relatively intact. Jackson accordingly maintained that there are "higher" and "lower" uses of language: the intellectual and emotional uses—or, as he later referred to them, the *superior* and *inferior* forms of speech. The former were genuinely expressive of thought; the latter were not. The intellectual, or superior, functions are the most labile, but even when severely impaired, some use of language and some residue of emotional expression are preserved. What these patterns of pathology suggested to

Jackson was the possibility that the same behavioral function was actually under multiple—and hierarchical—control. Some manifestations of the function could be maintained while others were destroyed. And some of the manifestations, he thought, required more development than others; for example, representational speech demanded more sophisticated development than did emotional speech. As he analyzed the syndromes, typically the least-developed mode of the function was maintained, while its most-developed or sophisticated use was lost.

Jackson's appeal to multiple control was influenced by two further sources: associationist psychology and evolutionary theory (see Smith 1982a). As we shall explain, the first source reinforced his repudiation of a localizationist program and his search for a different mode of explanation; the second provided a theoretical basis for his hierarchical model. Jackson derived both of these themes from Herbert Spencer's *Principles of Psychology* (1855, 1872).[3]

The associationist program in psychology assumed a variety of forms throughout its long history.[4] The crucial element in associationism was the idea that complex knowledge was built up from smaller units—ideas, in its classical formulations—through a limited number of general principles of association. Thus, for Locke, "simple ideas" become the atoms of knowledge, which can be held before the mind, compared, and compounded. In Hume's hands the principles of association were more clearly and rigorously circumscribed. He says in his *Treatise* (1888) that the "qualities, from which this association [of simple ideas] arises, and by which the mind is after this manner convey'd from one idea to another, are three, *viz.* RESEMBLANCE, CONTIGUITY in time or place, and CAUSE and EFFECT" (p. 11).

When applied in the neurological realm by Jackson, the associationist program mandated the abandonment of localization, rejecting the faculty psychology upon which the localizationist program of Gall, Bouillaud, and Broca had been built. According to the associationist program as Jackson inherited it, just as complex ideas are produced from simple ideas by the processes of association, so also complex mental operations must be compiled operations composed of simple sensory-motor associations. Any difference between higher and lower cognitive operations will involve simply more associations between lower-level sensory-motor processing, and not a unique cognitive faculty. The cognitive faculties posited by classical localizationist approaches could not be among the basic capacities, and localization would be fruitless (Richardson 1986a). "Will, memory, reason, and emotion," Jackson tells us, "are simply artificially distinguished aspects of one thing, a state of consciousness" (1884, p. 66).

Having abandoned a faculty psychology, Jackson was also committed to abandoning the localization of specific faculties. Whereas Broca had taken the "faculty of articulate language" to be one of many discrete and primitive mental functions, localizable in regions of the cerebral cortex, even as early as 1866 Jackson thought it "incredible that 'speech' can reside in any limited spot." No doubt this was in large part because he thought "the so-called 'faculty' of language has no existence" (Jackson 1866, p. 123). Associationism left no room for organology, and without organology there was no room for classical cerebral localization.

The evolutionary commitment provided the positive dimension to Jackson's account of neuropathologies. As he said at the outset of his seminal lectures, "We shall be very much helped in our investigations of diseases of the nervous system by considering them as reversals of evolution, that is, as dissolutions" (Jackson 1884, p. 45; cf. Jackson 1882). Evolution here was not the simple "descent with modification" defended by Darwin; rather, the evolutionary views of Herbert Spencer are what inspired Jackson. Unlike Darwin, Spencer embraced an orthogenetic, progressive view of the evolutionary process. Evolution was the expression of an inherent tendency toward development from lower to higher forms, or from less to more complex forms of organization. Evolution, according to this vision, was "a passage from the most simple to the most complex" and "from the automatic to the voluntary" (Jackson 1884, p. 46). The culmination and natural result of evolution was consciousness. Yet these later, higher forms were not to be constructed de novo; they were, rather, further modifications of the basic plan provided by the lower, less complex form.[5] This would typically involve the imposition of a control mechanism on the early products of evolution, allowing utilization of these devices for more complex functions.

When he turned to the pathological syndromes, Jackson proposed that the nervous system was organized in a hierarchical manner, with different levels representing different stages of evolutionary development.[6] The higher levels are more recent and are connected with volition and consciousness. The lower levels are more primitive and are connected with habit and reflex. Each higher level, Jackson thought, must work through the lower levels. Thus, in animals arising later in phylogeny, newly evolved neural structures would arise that would modify and regulate the performance of brain components that had emerged earlier. At the base of this hierarchy are parts of the nervous system that directly respond to sensory stimuli or control motor output. These basic mechanisms are specialized and task-specific, each representing some specific movement of a specific part of the body. Here is the only point in the hierarchy where we can find neural structures able to work independently to perform a function, and, hence, the only point where we might try to decompose the operation of the whole in terms of component operations. The kinds of operations performed by these components, though, are not the sort that would suffice for localizing *cognitive* capacities. These mechanisms, according to Jackson, only control specific *motor* movements. There are no discrete mechanisms responsible for coordinated actions, and no components for planning or reasoning about actions. At the middle level we have motor centers that are less specialized and less task-specific. Each center represents some complex movement, compounded of simpler movements represented directly by the lower level. These middle-level centers effect associations between lower-level components. These centers are less organized in that they are less automatic and more receptive to modification by experience. Part of what is critical about these systems, however, is that they do not work independently, not representing within themselves the information they require to perform their tasks. They achieve their effects by regulating, modifying, and integrating the operation of the lower-level components. Motor centers at the highest level each represent to some degree the entire body or movements of it. Their function is to coordinate complex movements. But, once again, these are not independent structures; they are mechanisms for regulating lower-level structures. Jackson summarizes this tripartite view and offers an anatomical interpretation:

> The lowest motor centres are the anterior horns of the spinal cord, and also the homologous nuclei for the motor cranial nerves higher up. . . . The lowest centres are the most simple and most organised centres; each represents some limited region of the body indirectly, but yet most nearly directly; they are representative. The middle motor centres are the convolutions making up Ferrier's motor region [just anterior to the central sulcus]. These are more complex and less organised, and represent wider regions of the body doubly indirectly; they are re-representative. The highest motor centres are convolutions in front of the so-called motor region. . . . [They] are the most complex and least organised centres, and represent widest regions (movements of all parts of the body) triply indirectly; they are re-re-representative. (1884, p. 53)

What higher units do, from Jackson's perspective, is coordinate what is directly represented in distinct lower-level components and regulate the activities of these components. This is what Jackson means by re-representing or re-re-representing what is already represented at the lower level.[7]

Schematically (Figure 9.1) we may conceive of $(S_{11}, S_{12}, \ldots, S_{18})$ as specialized organs at the lower level. Each controls a single movement, and loss of the organ results in paralysis of the corresponding part of the body. Loss of S_{11} would mean the loss of a specific behavior—perhaps the ability to move an arm, or to flex a finger. Similarly, $(S_{21}, S_{22}, \ldots, S_{24})$ occupy the second level and re-represent the movements represented at the lower level. These involve an intermediate level of integration and coordination. Each exerts some control over a variety of movements,

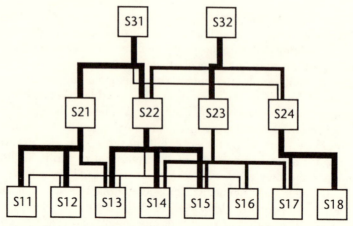

Figure 9.1. A Schematic Representation of a Control Hierarchy of the Sort Proposed by Hughlings Jackson (1884). Higher-level units exert a broader control, while lower-level units are more specific. Units at level 1 (S_{11}, . . . , S_{18}) control specific behaviors. Units at levels 2 (S_{21}, . . . , S_{24}) and 3 (S_{31}, S_{32}) coordinate the more specific behaviors of level 1. Breadth of the lines is meant to correspond to the strength of the connections; for example, S_{31} is strongly tied to S_{11} through S_{15}, but only weakly connected to S_{17} or S_{18}.

though the amount of causal influence they exert over units at the lower level is variable: S_{21} coordinates activity of three more specialized units (S_{11}, S_{12}, and S_{13}). Loss of S_{21} would entail a loss in the coordinated activity, but would not result in any paralysis, because each of these units can be activated by other higher-level units. Units at the highest level (S_{31} and S_{32}) produce the highest level of integration, but in so doing they exercise the least specific influence. They coordinate and integrate the activities controlled at lower levels; in Jackson's terms, they re-re-represent the movements represented at the lowest level and re-represented at the intermediate levels, and their degree of causal influence is also variable.

This hierarchical structure reflects our evolutionary heritage. By preserving the lower-level structure and function that was present in their evolutionary ancestors, higher, more developed organisms reflect the evolutionary history of the species; they recapitulate their evolutionary history. This preservation is most clearly revealed when the higher levels of the nervous system are destroyed or damaged. Jackson insists that symptoms will be both positive and negative. On the negative side there will be a loss of capacities: Some aphasics lose the ability to comprehend spoken speech; others lose the ability to speak. Epileptics may lose consciousness. This would mean that epileptics or aphasics would suffer a loss of the coordination and fine modulation of lower-level functions, but would not lose the most basic capacities. On the positive side, other neural centers will be expressed. Symptoms are a release of lower centers normally under higher-level control. Jackson insists, "Disease only produces negative mental symptoms answering to the dissolution. . . . [All] elaborate positive mental symptoms . . . are the outcome of activity of nervous elements untouched by any pathological process" (1884, p. 46). What would be left following the destruction of higher centers would be the more automatic, reflexive forms of behavior typical of lower organisms. Jackson refers to this process as the *dissolution* of the nervous system and claims to see it exhibited in the pathological syndromes such as aphasias. As evidence for this interpretation of the neuropathological syndromes, he points to the nature of the symptoms, which are typically negative. As we noted earlier, Broca's patient, Tan, was still capable of some vocalizations, but these were limited, simple, and automatic utterances such as oaths or simply "Tan." This is typical: voluntary applications of speech suffer more dramatically than do more automatic ones, and the latter suffer more than the applications used in emotional expression. With damage to the cerebral lobes, Jackson reasoned, the higher, voluntary uses of language would be lost and the lower functions would accordingly come to predominate. The patient would lose the ability to "convey propositions" through symbols, but would still have the ability to relate feelings.[8] This, from Jackson's perspective, reflects the maintenance of lower-level motor control despite the loss of higher levels of control over speech.

In one respect Jackson has simply repartitioned the nervous system into functional units, offering a different decomposition. His division into units crosscuts the division of the organologists. But the differences are more far-reaching than that would suggest. The localizationists divide the brain *horizontally* into a number of processing components, each operating at roughly the same level and in relative independence of one another. Each component has its own specific function. Jackson partitions the system *vertically*, with higher-level components operating on and modulating the behavior of the lower-level components. Only at the lowest level do we have specialized and independent modules. Moreover, the higher levels cut across the lower levels in their mode of operation, so we do not have a neat division of the cerebral lobes into regions responsible for regulating single, lower-level processes.

As we explained in Chapter 6, localization requires there to be a physical analysis corresponding to a projected cognitive organization. Even on Jackson's view, basic motor control is localized in this sense: there are physically discrete regions of the brain exercising limited control over specific behaviors. But Jackson's approach does not allow localization of cognitive capacities, and the more complex these capacities are the less localization will make sense. It is in this sense that his approach is non-localizationist. Higher-level, cognitive capacities do not have correspond-ing physical components with independent functions. Instead, they are attributes of the system because of its architecture and organization.

3. Parallel Distributed Processing and Cognition

Cognitive science is one area of research in which the decomposition of complex tasks into component tasks has been widely applied in recent decades. For the most part this has had a top-down orientation, with little attention to the details affecting realization. At its most extreme it has been antagonistic to physical details. The dominant metaphor is that cog-nition is information processing. The primary approach has been to ana-lyze complex information processing tasks into simpler information-proc-essing tasks and to seek simple mechanical realizations for the most basic tasks. As William Lycan, somewhat colorfully, portrays the program of research,

> We explain the successful activity of one homunculus . . . by positing *a team* consisting of several smaller, individually less talented and more specialized homunculi—and detailing the ways in which the team members cooperate in order to produce their joint or corporate output. (1987, p. 40)

In Lycan's hands it is homunculi as far down as psychology can see. Even the simplest component tasks in such cognitive theories remain ones of transforming information. In some accounts the information is understood as being represented in the system in symbolic structures. Processing is the transformation of symbolic structures into other symbolic structures; the symbols, in turn, are construed as semantically interpretable—that is, as referring to objects and having associated meanings (Fodor 1975). The resulting explanations describe the overall task in terms of operations that are intelligible given the semantic interpretation. For example, in a com-puter program that plays chess there may be operations that propose pos-sible moves, and other operations that project the resulting board posi-tions and evaluate them. The representations are of pieces and positions in a space defined by the board and the rules of chess. The operations are the legal moves. Since the symbols may refer to actual pieces and the operations to actual moves, we can readily understand how the program goes about playing the game.

This approach to cognitive behavior is commonly known as the *symbolic* approach, or, sometimes, as one relying on *rules and representations*. It is not the only possible strategy. When cognitive science was in its early infancy, another approach briefly emerged and then lay fallow. This com-petitor focused on networks composed of simple entities, supposedly sim-ilar to neural units, which exchanged activations and inhibitions (Rosenblat 1962; Selfridge & Neisser 1960). The goal of this *network* approach is to show how systems of simple entities can perform cognitive functions without operations on formally represented symbols. Some network models, such as Selfridge's Pandemonium, employed a homuncular representation of the overall task, decomposing it into significant subtasks. Other systems, such as Rosenblat's Perceptron, did not. A Perceptron takes in one pattern of activations and outputs another, with each input unit linked to an output by weighted connections. The network approach nearly disappeared with the publication of Minsky and Papert's *Perceptrons* (1969), which attempted to establish inherent limitations in network models. Recently, however, the network approach has reemerged and goes by such names as *connectionism, parallel distributed processing,* or *neural networks.* Whereas the symbolic approach epitomizes a variation on decomposition and localization, network models present a significant alternative program, proposing to explain cognitive functions without employing decomposition and localization.

It should help to develop the contrast more fully. The return to a cognitive psychology—that is, one acknowledging internal cognitive processes—after decades of dominance by behaviorism was inspired by two developments: the introduction of the digital computer, and Chomsky's (1957) defense of a generative grammar. The focus on formal systems in which symbolic structures are manipulated according to sets of rules is critical in both of these domains. Computers are often construed as list-processing machines, wherein lists of symbols coded in digital form are manipulated either by the hardware's configuration or by instructions stored in other lists of symbols that constitute the program. Chomsky's linguistics proposed to represent the set of grammatical sentences available in a natural language in terms of a finite structure. This would come in a set of rules that operate recursively on sequences of symbols, transforming these sentential representations while preserving grammaticality. Chomsky subsequently attempted to extend his analysis as a general account of human language. The basic idea has now been generalized into a wide-ranging set of models of human cognition in which the information to be processed is represented symbolically, and rules are introduced to manipulate these symbols. From our vantage point, what is important is that these rules and representations embody an attempt to account for an overall performance of the cognitive system by decomposing that task into simpler tasks.

One of the widely employed types of design for achieving cognitive performance within symbolic systems is the production system (Anderson 1983; Newell 1973). In production systems information is encoded symbolically in working memory (a variety of information can be encoded at any given time). The rules are conditionals: the *antecedent* specifies a test condition to be met by the symbols currently active in working memory, and the *consequent* stipulates an action to be performed. It is common to speak of the rule as *firing* if the antecedent is satisfied. This research program assumes that mental activity can be decomposed into a set of operations, each of which is governed by a set of rules operating on symbolic information. It has produced cognitive models successful in accounting for many features of human cognition; moreover, for many years it seemed the only option in developing an account of cognition.

A growing number of cognitive scientists have begun to explore the possibility that what *appears* to be rule-following behavior might only be *described* by these rules and that it might be feasible to explain cognitive performance without mirroring the rules in the mechanisms that explain that performance. The reemergence of network models, and the mounting evidence that more complex network architecture can perform tasks that previously seemed beyond the ability of the networks,[9] have led to a reexamination of the view that cognitive processes must be decomposed according to a rule. In this line of research, then, the decomposition of the symbolic approach is lost.

To illustrate why a network architecture is a significant alternative to the symbolic architecture, and how it promises to explain cognitive performance without engaging in the sort of decomposition that has been characteristic of the symbolic approach, it will be useful to describe briefly some common features of connectionist architecture and present two simple examples of connectionist systems. The basic components of a connectionist system are simple processing units able to assume varying states of activation and connections of varying strengths through which they can excite or inhibit others. There is tremendous variability in the network designs currently being considered, but what is common to all is that the units are simple and their interactions are critical. Activations can either be limited to discrete values (for example, 0 and 1), or can vary within a continuous range (say, from 0 to 1 or -1 to $+1$). Units can be connected with each other in various ways. They can be layered so that units within one layer project only to units in adjacent layers. Or, the network can be completely interconnected. Typically the input to a unit is determined by summing over the products of the activations of other units connected to it, and the weights associated with the connections between those units and the target unit. In Figure 9.2, the input to unit A is determined by multiplying the activation of B, C, and D by the weights connecting them to unit A, and then adding those values; this yields a net input to unit A of .24. The activation of A may then be determined by a variety of formulas. Some take the prior activation of A into account; some do not. The functions, moreover, may involve either linear or nonlinear functions in determining the activation level of A.

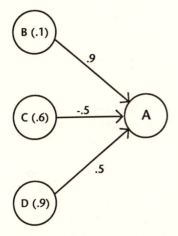

Figure 9.2. A Simple Connectionist System. In this case the input to A is an aggregative function of the activation levels of B, C, and D and their connection strengths to A.

One important source of variation arises in determining how the weights associated with the various connections are decided. They can be preset, but much of the interest in these networks stems from their ability to alter their own performance by changing the weights. This is done by using a variety of algorithms. Operations for changing these weights model learning in the networks; accordingly, the algorithms used are commonly referred to as *learning rules*. The simplest learning rules are variations on a suggestion by D. O. Hebb (1949). These are basic associationistic principles, requiring an increase in connection strength between units when they are simultaneously excited. For example, one variation might require increasing the strength of the connection between units in proportion to the product of their activation levels.[10]

In addition to determining the actual mechanics of a network, a researcher must also specify how these networks are to be construed as performing a cognitive function. This is usually spoken of as providing an *interpretation* for the activities of the network. There are two general approaches to interpreting the activity of such systems. The simplest is to let each unit represent a hypothesis or a goal. Since each unit has its own representational function, this approach is referred to as *localist*. The more complex, but potentially more interesting, interpretation is to treat a particular pattern of activation over an ensemble of units as serving the representational role. In such systems one pattern of activation over a set of units may receive one interpretation, while another pattern over the *same* set of units receives another. Since it is the pattern of activations that determines the interpretation, this approach is referred to as *distributed*.

The crucial move common to all network models is that of explaining cognitive performance without casting the explanation in terms of rules that manipulate symbols. Rules and representations are both semantically interpretable structures; hence, explanations of the constituent operations employ a framework that makes sense from the perspective of the overall cognitive task. Network models do account for the cognitive performance, but often they do so without providing an explanation of component operations that is intelligible in terms of the overall task being performed. The network is a cognitive system; the components are not. The result is that we do not explain how the overall system achieves it performance by decomposing the overall task into subtasks, or by localizing cognitive subtasks.

In order to show how network models provide an explanation that does not involve decomposing the overall task, we will briefly describe two simulations. The first involves a two-layer network learning to recognize patterns.[11] The overall process is one that could figure either in basic perception or in categorizing objects already perceived. The network consists of eight input and eight output units, with each input unit connected to each output unit (see Figure 9.3). The activation (a_j) of an output unit j is the sum over the products of the input activations for each of the i units, and the weights of the connections linking them to the output units (w_{ji}): $a_j = \Sigma_i a_i w_{ji}$. The input arrays can be viewed as representing objects belonging to four different categories. We will suppose arbitrarily that these are cup, bucket, hat, and shoe. Table 9.1 shows the input patterns that correspond to a prototypical instance of each category, and the target output patterns that the network is trained to approximate. The target outputs can be thought of as the system's names for the four categories. For example, the input array for a prototypical bucket is $\langle -1, -1, +1, +1, +1, -1, -1, -1 \rangle$, and the output array for "bucket" would be $\langle -1, -1, -1, -1, +1, +1, +1, +1 \rangle$.

The goal of the network is to learn the association between the input array and the "name." In training the network, the actual inputs and target outputs were distorted by a randomly chosen amount between -0.5 and 0.5 to capture the fact that we do not always encounter prototypical objects or undistorted names. Thus, where the pattern designated for the prototypical cup is $\langle -1, -1, -1, -1, +1, +1, -1, -1 \rangle$, an actual input on one trial might be $\langle -0.76, -0.89, -1.21, -1.01, +1.33, +0.99, -0.65, -0.92 \rangle$. The network was trained over fifty epochs, or training sets. During each epoch the network received a distorted version of each input and its corresponding distorted target output. The network used the *delta rule*, which adjusts the weights (w_{ji}) of each connection leading to each output unit on each trial by an amount proportional to the product of the difference between the actual output (a_j) and the target output (t_j) and the activation of the relevant input unit (a_i) on that trial. This can be represented as $\Delta w_{ji} = .0125 (t_j - a_j) a_i$. If the target output for U_5 is -1.0, and

Object	Prototypical Inputs								Target Outputs							
	I_1	I_2	I_3	I_4	I_5	I_6	I_7	I_8	U_1	U_2	U_3	U_4	U_5	U_6	U_7	U_8
Cup	−1	−1	−1	−1	+1	+1	−1	−1	−1	−1	−1	−1	−1	−1	−1	−1
Bucket	−1	−1	+1	+1	+1	−1	−1	−1	−1	−1	−1	−1	+1	+1	+1	+1
Hat	−1	+1	+1	+1	−1	+1	−1	+1	−1	+1	−1	+1	−1	+1	−1	+1
Shoe	+1	+1	+1	+1	+1	+1	+1	+1	+1	+1	+1	+1	+1	+1	+1	+1

Table 9.1. Prototypical Inputs and Outputs for Two-Layer Pattern-Recognition Network with Eight Units per Layer.

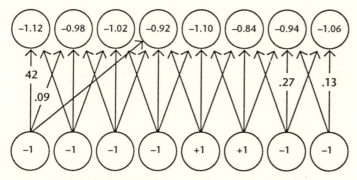

Figure 9.3. A Two-Layer Pattern-Recognition Network. All units at the lower level are connected to each unit at the higher level, with no intervening levels. For simplicity, not all connections or connection weights are shown.

its actual output is −0.8, then the above input array would adjust the weights of the connections to U5 from the input units up or down by the following amounts: ⟨+.00190, +.00223, +.00303, +.00253, −.00333, −.00248, +.00163, +.00230⟩. In this instance the rule increases the strength of the connection weights from input units having the same sign as the target output, and decreases the rest.

After training through the full fifty epochs, the network was tested on three different types of input: the actual prototype of each category, an instance randomly distorted in the way described above, and an input for which the sign of one of the input units of the prototype was reversed. The test inputs are detailed in Table 9.2. When presented with an actual prototype, or with a distorted version of the input, the outputs were all within .5 of the target output. Even when given a pattern in which one of the eight input values was reversed in sign from the prototype, the system produced outputs that were positive or negative as appropriate in all but one case.

Object	Test inputs								Outputs							
	I_1	I_2	I_3	I_4	I_5	I_6	I_7	I_8	U_1	U_2	U_3	U_4	U_5	U_6	U_7	U_8
Cup	-1.00	-1.00	-1.00	-1.00	+1.00	+1.00	-1.00	-1.00	1.12	-0.98	-1.02	-0.92	-1.10	-0.84	-0.94	-1.06
	-0.76	-0.51	-0.82	-1.11	1.47	0.82	-0.83	-0.90	-0.81	-0.90	-0.71	-0.83	-0.77	-0.72	-0.62	-0.89
	-1.00	-1.00	-1.00	-1.00	+1.00	*-1.00*	-1.00	-1.00	-0.86	-1.39	-0.85	-1.41	-0.26	-0.78	-0.16	-0.89
Bucket	-1.00	-1.00	+1.00	+1.00	+1.00	-1.00	-1.00	-1.00	-0.99	-1.06	-0.98	-0.96	0.91	0.94	0.99	0.88
	-1.00	-0.54	1.34	0.63	0.98	-0.59	-1.24	-0.81	-1.06	-0.81	-1.03	-0.68	0.63	1.00	0.70	0.88
	-1.00	-1.00	*-1.00*	+1.00	+1.00	-1.00	-1.00	-1.00	-0.98	-1.24	-0.96	-1.22	0.30	0.06	0.39	**-0.03**
Hat	-1.00	+1.00	+1.00	+1.00	-1.00	+1.00	-1.00	+1.00	-0.91	0.96	-0.87	1.05	-0.84	1.06	-0.90	0.92
	-1.18	0.62	1.20	0.87	-1.21	1.38	-1.02	1.48	-1.07	1.11	-1.01	1.22	-1.12	1.10	-1.18	0.92
	-1.00	*-1.00*	+1.00	+1.00	-1.00	+1.00	-1.00	+1.00	-1.20	0.38	-1.14	0.49	-0.74	0.87	-0.75	0.68
Shoe	+1.00	+1.00	+1.00	+1.00	+1.00	+1.00	+1.00	+1.00	0.99	0.94	1.05	1.07	0.93	1.03	0.92	1.15
	1.42	1.44	0.64	1.31	0.72	1.24	1.03	1.19	1.20	1.28	1.25	1.39	0.81	1.00	0.77	1.15
	-1.00	+1.00	+1.00	+1.00	+1.00	+1.00	+1.00	+1.00	0.13	0.75	0.21	0.85	0.38	1.18	0.41	1.15

Table 9.2. Test inputs and corresponding outputs following 50 training epochs, based on target outputs in Table 9.1. In the test phase, the system was presented with three test inputs for each of the four items. The first row for each item gives the prototypical input and the actual output. The second row shows randomly permuted prototypes and the responses to them. And the third row gives the prototype, but with the sign of one input unit reversed (indicated by italics). With one exception (indicated in boldface), the outputs are at least of the right sign.

This two-layer network is very simple, but it does succeed in recognizing and classifying the patterns. In general, this sort of pattern recognition is something that connectionist networks do well; it is also something quite difficult for symbolic systems. Consider how one would design a symbolic system to perform this task. The input pattern for an object is simply an arbitrary array of activations produced in units of the system. In more realistic situations, this array would be produced by a set of feature detectors in the visual system, and the values in the array would represent values on these features. For example, a positive value on an input unit might indicate the presence of a feature or the activation (confidence) level of the corresponding feature detector. The values of the output units would likewise represent features of either the word or the mental representation of the object. A symbolic system would begin with encodings of the features of the input and seek to develop rules that would produce the symbolic representation of the output. The rules would have to specify what set of features would constitute, for example, a shoe. Since there is significant variability between shoes, the rules would have to specify the various combinations of features in the input that should still result in recognizing the object as such. The attempt to produce such rules has overwhelmed AI investigators.[12]

A connectionist approach, by contrast, does not set out to identify rules. Rather, the connections in the network are allowed to adjust during the training phase until the network can efficiently distinguish the objects in the domain. For the network to accomplish this, it must have structure. This is found in the connections, which serve the function of rules in a symbolic system. We sometimes can interpret the connections as providing rules for how to identify, say, balls as opposed to shoes on the basis of features. It is important to recognize how these rules are obtained. They are *not* developed by specifying the conditions under which an object of a given sort is present. Instead, the rules result from the network's discovering the correlation between features and objects. Each weight represents the reliability of a specific feature as an indicator within the class of objects. Thus, in the array given in Table 9.1, a positive value for I_7 is a good indicator for a shoe, while a positive value for I_2 tells us less (since it is ambiguous between hats and shoes). The learning rule allows the system to adjust dynamically to find a set of weights that achieves the best fit, given the class of inputs. What is important for understanding the contrast is that, except in the limiting case in which the weight to an output unit is 0, each input feature contributes something to the net output, and generally none is individually sufficient to determine the output. Thus, no decomposition into meaningful subtasks is needed.

Two-layer networks of the sort we have just described are able to learn to compute many relations between inputs and outputs, but there are limits to what they can learn. Such limits are the focus of Minsky and Papert's (1969) criticism of Rosenblat's Perceptron model. Minsky and Papert show that while a two-layer network can compute functions like AND and inclusive OR, it cannot compute a function such as exclusive or (XOR). A two-layer network is unable to compute a function for which the target outputs cannot be linearly separated.

Two steps are required to surmount this sort of difficulty. The first is to insert one or more layers of units between the input and output layers. The second is to use a nonlinear activation function such as the logistic function $a_i = 1 / (1 + e^{-net(i)})$, where $net(i)$ is the sum of the products of input activations and connection weights. The second network we will describe uses multiple layers and the logistic activation function. Our interest in multilayer networks, however, is not due simply to their greater computational power. The intermediate layers of units—generally referred to as *hidden units*—can be viewed as performing component processing steps: they process information received from earlier layers and then provide this new information to units in later layers for further processing. But if the network can learn, then the determination of the computations performed by the hidden units, as well as what the hidden units represent, are dynamically determined by the system. Moreover, the information represented by these units does not precisely correspond to the type of information that would normally be employed in a symbolic decomposition of the overall task.

This is clearly seen in a network designed by Hinton (1986), which learns information about the two isomorphic family trees, one English and one Italian, shown in Figure 9.4. The information in these trees can be encoded in 104 simple relational propositions of the form $\langle person_1$ R $person_2\rangle$ (for example, Colin has mother Victoria). Hinton constructed a network containing 36 input units and 24 output units. Twenty-four of the input units stand for the individuals in the two families and are the values that can be assigned for "$person_1$." The other 12 input units give the possible values for R. The twenty-four output units also represent the individuals, though this time as the possible values for "$person_2$." In Figure 9.5, modified from Hinton (ibid.), we show the two sets of units representing individuals (*P1* through *P24*) and the twelve indicating geneological relations (*R1* through *R12*). Because a single unit stands for a person or a relation, this is an instance of a local representation. In this system, in addition, there are three layers of hidden units. The first layer consists of 12 units, 6 of which (*H1* through *H6*) receive inputs from the 24 units coding for $person_1$, while the other 6 (*H7* through *H12*) receive inputs from the 12 units specifying the relationship. The second layer also consists of 12 units, which receive inputs from all 12 units in the first hidden

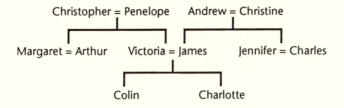

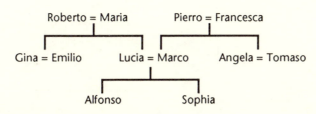

Figure 9.4. Two Isomorphic Family Trees. Hinton (1986) examines a connectionist network designed to deal with information concerning family relationships learned from the two isomorphic family trees depicted here.

layer. The final layer of hidden units consists of 6 units. The fact that the 24 units coding for person$_1$ must feed their information through a bottleneck of 6 hidden units forces the network to find a distributed representation of the different individuals that captures whatever information about the individuals the network requires to complete its task. The same principle operates with respect to the 6 hidden units receiving input from the 12 units specifying the relationship.

The network was trained to identify the correct person$_2$ when given person$_1$, and the relationship for 100 of the 104 relational propositions using the back-propagation algorithm as the learning rule.[13] After 1500 cycles of training,[14] Hinton's network not only learned to complete all 100 training propositions, but was also able to generalize to the four remaining propositions.[15] How the network accomplished this is significant. We can determine the representational function assumed by the first set of hidden units (6 are connected to the person$_1$ units, and 6 to the relationship units) by examining the weights between the inputs and this layer. In Figure 9.6 the 6 hidden person$_1$ units and the weights coming into them from the 24 person units are displayed. White boxes indicate positive weights on a connection; black boxes indicate negative weights. The size of the boxes represents the strengths of these connections. It is clear that the hidden units have extracted useful information about the individuals, despite the fact this information was never explicitly provided to the network. Thus,

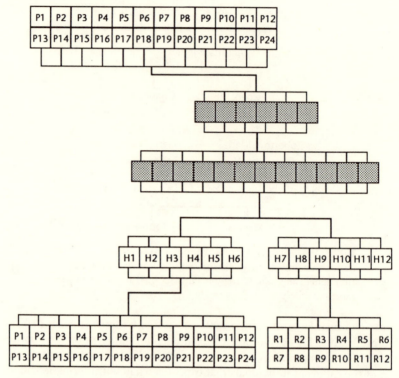

Figure 9.5. Five-Layer Network. A schematic representation of the network used by Hinton, with three layers of hidden units. P1 through P24 represent the individuals in Figure 9.4. R1 through R12 represent geneological relationships. The existence of a bottleneck forces the network to utilize a distributed representation.

unit 1 and (to a lesser degree) unit 5 identify family membership, while the remaining units are all generally indifferent to family memberships. Units 2 and 3 appear to encode the individual's generation (unit 2 responds most positively to older members, unit 3 to younger members). And units 4 and 6 seem to be representing membership in the two root families in each tree (unit 4 favoring the right side of the two trees, unit 6 the left).

Two comments are in order. First, it is the network, not the theorist, that determines what information to represent in the hidden units. Second, while it is often possible, as in this case, to label the hidden units in terms of what they represent, these labels are approximate; typically, it is difficult, and sometimes impossible, to fix these labels. For example, if unit 1 does indeed represent the English branch and unit 5 the Italian, they do not treat all members of the respective families equally. Penelope

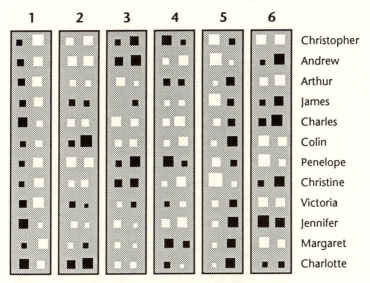

Figure 9.6. Six Hidden Units with Inputs from Twenty-Four-Person Units. Each node in the 6 hidden units represents one of the 24 individuals. The right-hand column in each of 1 through 6 represents a British individual, as indicated. The left-hand column within each represents the corresponding Italian family member (see Fig. 6.4). White boxes indicate a positive weight, so that, for example, inputs from any British name will tend to produce positive activation on all the units within the right-hand column of unit 1, while inputs from an Italian name will not. Black boxes indicate a negative weight, and box size indicates the level of activation. (From Hinton 1986.)

is treated by unit 1 as more of an English person than Charles, and Francesca is treated by unit 5 as more Italian than Emilio. In the six units encoding relationships (see Figure 9.7), unit 10 seems to represent the sex appropriate for the target person, but for others, such as units 9 and 12, it is difficult to specify what information is represented. Thus, while sometimes we are able to assign labels to the information-processing activities of the hidden units, these labels are partial and approximate. Moreover, the particular information-processing task a unit carries out may not be one we can describe at all. Thus, while we can construe multiple-layer networks as decomposing the information-processing task, the decomposition is not one advanced by the theorist and perhaps not even one the theorist can describe, at least not in the vocabulary in which the overall task is defined.

Connectionist models explain performance without explicitly or necessarily decomposing that performance into intelligible subtasks. Insofar as cognitive scientists accept connectionist simulations as explaining cogni-

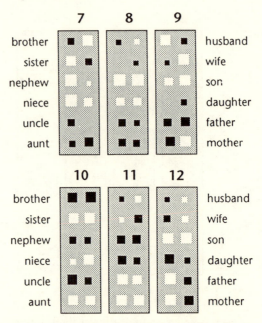

Figure 9.7. Units Recording Relationships and the Apparent Representation from Hidden Units. Projection from the 12 input units that represent relationships to the 6 in the second layer. In this case the representational functions are far less evident. (Also from Hinton 1986.)

tive performance, they are making a significant break with the decompositional strategy of traditional mechanism. It is still too early to determine how successful the connectionist strategy will be. It may be that connectionist approaches will only be useful for modeling low-level cognitive tasks such as visual perception and will fail in domains of reasoning and linguistic performance (cf. Fodor and Pylyshyn 1988; Pinker and Prince 1988). It is not important that we take a stand on the ultimate viability of connectionism as a framework for cognitive theorizing in order to make our main point: connectionism represents a break with traditional mechanism, pointing toward a different category of models and employing an alternative strategy for developing them. This alternative emphasizes systems whose dynamic behavior corresponds to the activity we want to explain, but in which the components of the system do not perform recognizable subtasks of the overall task. As with the case of Jackson's model, the decomposition of the system fails to correspond to cognitive organization. The overall architecture of the system—and especially the way components are connected—is what explains cognitive capacities, and not the specific tasks performed by the components. We have abandoned decomposition and localization.

4. DISTRIBUTED MECHANISMS FOR GENOMIC REGULATION

Stuart Kauffman (1986; forthcoming, chs. 9–13) has developed a network model for examining the structure and origin of genomic regulatory systems. His work can be understood as directed, in part at least, to the question of how genetic regulatory systems can be maintained in the face of genetic mutation and recombination.[16] One result of research on the mechanisms of gene expression in recent decades has been the recognition of complex sets of genes that regulate the expression of other genes. The activity of regulatory genes is probably relatively specific, but what is indicated by this work is a fairly complex network of genes connected either directly or indirectly; as a consequence, any mutations or transpositions affecting regulatory genes would alter the expression of the genetic system as much as—indeed, more than—would mutations in nonregulatory genes. The changes in these genes should have complex ramifications downstream. The more complex and interactive the regulatory system is, the more unstable and delicate it would seem to be, as it would offer more loci where the effects of mutation could have widespread effects.

Kauffman claims that unrealistically high selection pressures would be required to counter the mutation rates and maintain a regulatory system, once it becomes sufficiently complex. To give a definite form to the claim, he develops models with an arbitrarily defined set of "correct" connections and takes fitness to be a function of the frequency of such connections. If we assume a fixed mutation rate and a basal fitness of 0, then the class of fitness functions is given by $W_x = (G_x/T)^a_1$, where T is the total number of regulatory connections in the network, and G_x is the number of the "correct," or "good," connections. The value of a corresponds to three ways fitness can vary with changes in the frequency of good connections: If $a = 1$, then fitness falls off linearly as the frequency of bad connections increases. If $a > 1$, the fitness function is concave, falling off steeply at first and then leveling off. If $a < 1$, the fitness function is convex, falling off slowly at first and then more rapidly. If we suppose the mutation rate is constant, then, Kauffman reasons, because fitness is inversely proportional to T, increasing the number of regulatory connections will mean a decrease in the significance of selection; that is, increasing T will decrease the absolute fitnesses of alternatives and thereby decrease their absolute-fitness differences.[17] Mutation pressures will then be more likely to overcome differences in fitness. As complexity, measured by T, increases, mutation is more able to drive networks off the adaptive peaks.

The net result of Kauffman's work is to change the question: Whereas someone with a more conventional viewpoint would perceive the complexity as an adaptation maintained in the face of selection and would confront the problem of identifying suitable forces, from Kauffman's vantage point the stability of the regulatory system may not be something that requires a special explanation in terms of selection. Kauffman suggests that, in fact, the genome is spontaneously self-organizing. To support this suggestion he again examines statistical features of systems with multiple interconnections. He shifts from seeking a highly localized explanatory unit to exploring the power and structure of a distributed system. Decomposition and localization would assume that one could identify discrete genetic units responsible for specific characteristics of the system, including regulatory control. Accordingly, some researchers have tried to identify specific connections between regulator genes and the genes regulated. Kauffman looks instead at the patterns of interactions between genes, not at the specific effects of any genes. In many cases, he says, the self-organizing pattern of connections in the genome may be the critical determinant of the genome's behavior, rather than any gene or set of genes that must be maintained by selection.

Kauffman's model networks have a fixed number of "genes," treated as nodes in the model, with a set number of random connections between them. These connections provide the vehicle for one gene to regulate the behavior of another. Kauffman then analyzes the properties of such networks. The ratio of the number M of connections to the number N of genes affects the expected number of genes any one gene potentially influences directly. This in turn affects (1) how many steps are required for a gene to communicate to all that it regulates and (2) the chances of a given gene receiving feedback, via a loop, from itself. In networks where M exceeds N, connected circuits typically form: some genes have regulatory influence on most others, many genes lie on feedback-loops and eventually receive activation from themselves. Just beyond the point where the ratio of M to N exceeds 1.0, both the average radius of effect of each such gene and the mean length of feedback-loops between genes increase. If M then becomes much larger than N, however, both the average radius and mean length of feedback-loops fall as more and more genes are connected directly.

Kauffman's model regulatory networks (forthcoming, ch. 5) rely on simple Boolean operations. That is, he confines himself to networks where each unit (gene) is limited to one of two values (on, off) and in which there are deterministic transitions according to Boolean operations. He focuses on systems in which the value of one unit suffices to guarantee that another unit assumes one of its two values. Kauffman refers to such systems as *canalizing* and says that the majority of known genes in bacteria and phage are governed by such Boolean functions (1986, p. 174).[18] He then shows that even when such systems are confronted with random alterations—that is, mutations—they will exhibit highly ordered regulatory behaviors. There will be a spontaneous, natural order to the system.

As a simple illustration, consider the three-unit system depicted in Figure 9.8: each unit sends activation to the other two, and the response of each to the incoming signals is a Boolean AND or OR operation. The value of unit 1 is + if the values of both units 2 and 3 were + on the previous cycle; otherwise it assumes a − value. Unit 1 is governed by an AND function. Units 2 and 3 will record a + value if any other unit assumed a + value in the previous cycle; otherwise they will assume a − value. Units 2 and 3 are governed by an OR function. Both functions are canalyzing, as unit 1 will go to − if either unit 2 or unit 3 is − on the previous cycle, and units 2 and 3 will go to + if either of the other units are + on the previous cycle. A system with three Boolean nodes has eight possible patterns of activation. Assuming the network is synchronously updated—that is, a pattern at time t completely determines the new pattern to be found at $t + 1$—then, because there are a finite number of states, and deterministic transitions, the system will inevitably encounter cycles. Once it enters a cycle, it will repeat it indefinitely. These cycles are what Kauffman calls *dynamical attractors* or *attractor state cycles*. For example, the alternation between $\langle - - + \rangle$ and $\langle - + - \rangle$ is a stable cycle. Which cycle a given network will settle into depends totally upon the starting point.

Kauffman suggests that Boolean nets have a number of properties that make them suggestive for understanding genomic regulation. If we limit our attention to cases in which $M/N > 1$ and to relatively large systems in which N is on the order of 10^4 or 10^5, then these cases will also find dynamical attractors. It turns out that such systems have parallels with the behavior of eukaryotic cells. For example, the number of cell types in an organism and the number of attractor state cycles in a network are both roughly the square root of the number of constituent genes. More importantly, both turn out to be very stable in the face of mutations. Kauffman has carried out computer simulations in which mutations occur as random alterations in the wiring diagrams of a population of 100 networks. Given appropriate values for M and N, the stability in the model networks is so great that 90% of possible perturbations in these networks leave no net change on the overall stable state of the network. Even when units are deleted from the network, only 10–15% of other units alter their pattern

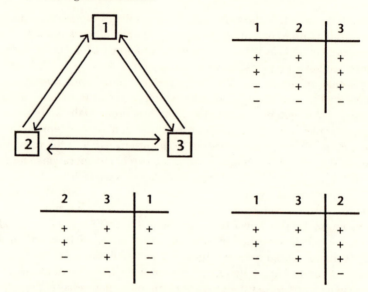

1	2	3
+	+	+
+	−	+
−	+	+
−	−	−

2	3	1
+	+	+
+	−	−
−	+	−
−	−	−

1	3	2
+	+	+
+	−	+
−	+	+
−	−	−

Figure 9.8. A Simple Boolean Network with Three Nodes. In this case we have three nodes influencing each other. Unit 1 computes an "and" function, assuming a + value at step n+1 if and only if both units 2 and 3 assume a + value at step n. Units 2 and 3 compute "or" functions, assuming a + value at step n+1 if and only if at least one of the other two units assume a + value at step n. The system will exhibit simple cycles, or attractors; for example, both the case where all units are on and the case where all are off are stable, or attractor, states.

of expression. The reason for the stability of such systems and their ability to migrate to only a few alternative states is that a majority of the units in the system settle into fixed activation states that do not alter as the system cycles, thereby producing large *forcing structures* which inhibit propagation from other units through the system. Only the remaining isolated networks are capable of undergoing change, so they alone determine the range of variability in the system.

If the genome consists of large ensembles of genes mutually influencing and regulating one another, then there will be a stable, inherent order to the system independent of the influence of selection. The networks of genes will tend to be stable, independently of selection. As Kauffman puts it, given "statistical theories of the expected structure and behavior" of genomic regulatory networks, then those aspects of structure and behavior that would be predicted constitute "typical or generic properties of the ensemble of genomic regulatory networks" (forthcoming, ch. 11).

Kauffman offers his models as useful characterizations of the null state against which selection must operate. He contends that without a characterization of the null state, evolutionists are prone to attribute too much power to natural selection, viewing it as capable of generating almost any possible genetic state that is highly adaptive. In contrast, Kauffman contends that the null state has quite powerful properties which selection is generally powerless to overcome:

> A general implication of [this class of models] is that a sufficiently complex genetic regulatory wiring diagram will approach arbitrarily close to the typical organizational properties of the unselected system. Thus, for sufficiently complex genomic systems, predications from the typical properties to be expected in the absence of further selection to those actually found in organisms would be reasonably accurate. (1986, p. 180)

Kaufmann's vision of the genetic regulatory system is close in spirit and method to connectionist models of psychological functioning. In both, it is ultimately the statistical pattern of the connections in the system, and not the jobs performed by specific units in the system or outside the system, that is critical to the behavior of the system. One respect in which the case examined by Kauffman is interestingly different from that discussed by connectionists is that, on the basis of his network model of the genetic regulatory system, Kauffman argues for a need to revise the question of what needs to be explained. What *appeared to* require explanation was the ability of selection to maintain the regulatory system. If Kauffman is correct, however, this will turn out not to be an issue. To the degree the regulatory system is stable, it is so *independently* of selection; the stability turns out to be due to self-organizing features of the genetic network itself. What requires explanation through selection is not how the system maintains its stability, but how it can be transformed from the stable state.

5. Conclusion: Mechanistic Explanations without Functional Decomposition and Localization

We have briefly described three cases in which researchers have pushed beyond classical mechanistic views. In Part III we saw that traditional mechanism, guided by localization and decomposition, explains why a system behaves as it does in terms of the behavior of its individual components. Even the explanations of complex systems with a variety of integrated circuits still make a major appeal to the contributions of specific modules in the system. In the cases sketched in this chapter, by contrast, this strategy is abandoned; instead, the approaches attempt to show how the properties of the system emerge simply as a result of the connectivity of very simple components in a network. In the most extreme cases the parts do not individually perform any activities that can be characterized in terms of what the system does. Decomposition and localization fail. The surprising feature of these networks is that the pattern of connections results in systemic properties that would not be anticipated by focusing on the contributions of component units.

While network models are not classical mechanistic models, there is still a clear sense in which they are mechanistic. The behavior of the system is a product of the activities occurring within it. All the components are simple mechanical units, and their interactions are all characterized in simple mechanical terms. If the models are well motivated, then component function will at least be consistent with physical constraints. The difference is that what is important in determining the behavior of the system in a network model is not the contribution of the parts, but their organization. In simpler network models the parts are interchangeable; indeed, they are typically simple on/off units. Their role is to excite or inhibit the activities of other units in the system. The connections within the system determine the patterns of behavior that are observed in the system. There is clearly no case here for abandoning a mechanistic perspective. Nonetheless, these systems defy the approach to mechanism that we charted in earlier chapters, because these systems are neither decomposable nor even minimally decomposable, and systemic functions cannot be localized. Whether we are interested in cognitive capacities or genomic regulation, analyzing the components of the system in isolation throws no light on the phenomenon under investigation. One can only produce the phenomenon in the whole (or nearly whole) system. Analytic techniques that focus on the behavior of individual components through excitatory or inhibitory studies will fail; moreover, synthetic approaches are not liable to reveal component structure or organization. As a consequence, localization and decomposition break down with network systems.

Connectionist systems thus defy some of our traditional tools for studying natural systems, for these tools rely on being able to decompose the system, work on components singly, and then build up again to understand the whole. It should not surprise us that there are such connectionist systems when we recall the fallibility of heuristics. For many problems localization and decomposition have been highly successful, either in giving accurate accounts of how certain systems work, or accounts that constitute good approximations. When natural systems are not nearly, or at least minimally, decomposable, then those heuristics will lead us astray. Human cognition and genetic regulation may turn out to be processes for which localization and decomposition fail. Insofar as connectionist systems are intrinsically parallel systems, it is not possible to understand them by looking at changes in one component at a time. And while it may be true that one can always (in principle) simulate the behavior of a parallel system serially, characterizing the transformations the system undergoes, the operating principles and architecture are fundamentally different. To understand the behavior of these systems it is necessary to understand that a multitude of such changes occur simultaneously.[19]

The development of network models is likely to alter our conception of machines and mechanism in radical ways. In particular, it may alter our understanding of the respects in which properties of a mechanistic system can be said to be *emergent* (cf. Bechtel and Richardson 1992). Advocates of emergentism have maintained that certain kinds of systems are capable of giving rise to radically new properties not present in the components of the system. Such appeals to emergence have struck many as mysterious, others as trivial. On the one hand, it is too easy to see that when a certain degree of complexity is reached in a system, new properties will appear: a square has properties none of the component line-segments have. On the other hand, emergence is vacuous unless we have some account of how the organization matters. In network systems we can understand how emergent properties appear without waxing mysterious. In calling the systemic properties of network systems emergent, we mark a departure from the behavior of simpler systems and indicate that traditional mechanistic strategies for understanding network systems may simply fail. But the behavior of the system is not unintelligible or magical; it follows from the nature of the connections between the components within the system.

The emergent behavior of connectionist systems may be mysterious in another sense, however: We may not be able to follow the processes through the multitude of connections in a more complex system, or to see how they give rise to the behavior of the system. We may fail in the attempt to understand such systems in an intuitive way. To quote Simon once again, complex systems that are not hierarchical and decomposable "may to a considerable extent escape our observation and understanding" (1969, p. 219).

CHAPTER 10

Constructing Causal Explanations

1. Decomposition and Localization in Perspective

Our focus has been on decomposition and localization and their role in the development of scientific research programs. We have especially emphasized their heuristic role in the construction of explanatory models within problem domains that are ill structured. In the cases we have discussed there was initially no well-delineated space of explanatory models, nor even a clearly defined range of phenomena to be explained. The problem-solving tasks were, then, exploring and constructing the space of explanatory models, and determining the precise range of phenomena to be explained. These tasks are interdependent, their relationship neither simple nor uniform. As Ian Hacking (1983) observes, the relationship between theory and experimental "data" can change with the development of a theory and varies from one science to another. Observations may sometimes be undertaken with no particular theoretical motivation; however, the phenomena one then finds may have no clear significance, for lack of a theory to make sense of them. Alternatively, the observations' significance may become clear only after the fact. In other cases, a program of experimental investigation and observation is undertaken with clear theoretical motivations. In these instances the implications may be clear and immediate. However, on occasion the significance changes and is far from what the initial experimenter thought it would be.

It would be a mistake, from our current vantage point, to try to fit the full range of cases into a single mold. In the cases we have discussed, research programs vary greatly in both motive and method. Some motives were clearly practical: the investigation of fermentation, for example, mattered for the development of wine and beer industries, and research in genetics was not unrelated to agriculture. Other motives were more theoretical: the importance of catalytic reactions to respiration could not be clear in the absence of Lavoisier's view that respiration was a slow combustion. Even the relationship of fermentation and respiration was unclear. At times it was assumed that there should be a single, general theory encompassing both; at other times they were treated in relative isolation. Likewise, it was not always clear what the significance of linguistic or cognitive deficits was for a theory of neural and cognitive functioning. Gall and Spurzheim rejected appeals to deficits on the grounds that the resulting phenomena were unreliable and of unclear relevance to normal functioning; Broca and Wernicke made them a centerpiece of their work; Jackson acknowledged their importance, but drastically reinterpreted them; and in contemporary research their significance is still a matter of debate. In part, decomposition and localization suggest a form for explanatory models; in part, they serve to impose a structure on the phenomena to be explained. Exactly how they affect the development of explanatory models depends on the theoretical context and the available experimental techniques.

We have described the explanatory models generated by these heuristics as mechanistic in character. This is meant to suggest a number of features common to the cases we have discussed. First of all, the models do feature in what are naturally thought of as *causal explanations*. Models describing the structure of the genetic material are meant to explain development and inheritance in terms of underlying causes. For example, the presence or absence of specific enzymes explains respiration in these terms. However, laws and theories classically construed have little place in the picture. The problems guiding the sort of research we have recounted are (1) explaining how some particular effect is actually produced, here and now, and (2) by what means. The resulting explanations are sometimes what Nancy Cartwright (1983), following John Stuart Mill (1884), calls "explanation by composition of causes." She says this "picture of how nature operates to produce the subtle and complicated effects we see around us is reflected in the explanations that we give: we explain complex phenomena by reducing them to their more simple components" (1983, p. 58). We describe the effect on a body, say, as the consequence of gravitational and electrical forces acting together. Likewise, we describe our linguistic competences in terms of the effect of a multitude of capacities, and we explain phenotypic traits as the consequence of the influence of genetic and environmental contributions. We explain a phenomenon that interests us by identifying a variety of causal factors and showing how they conspire to yield an effect.

Second, the explanatory models appeal to what are naturally thought of as underlying mechanisms. Fermentation is something done by cells; the mechanism is a complex biochemical pathway. Development is something organisms undergo; the mechanism that regulates it is a complex genetic pathway. Language or cognition is something attributable to people; the mechanism is given in neurophysiological—or, perhaps, functional—terms. Thus, the models we have looked at are tied to more than one explanatory level. We shift down from the system to its parts in order to explain how the system does what it does. The models are related to what Darden and Maull (1977) call *interfield theories* (see also Darden 1986). A theory of fermentation must explain fermentation in terms of the biochemical mechanisms that effect it; a theory of development must explain development in terms of the genetic or epigenetic mechanisms that regulate it; and a theory of cognition must explain cognition in terms of the mechanisms, whether psychological or neurophysiological, that mediate behavior. This has important implications for the dynamics of the resulting theories.

We have also focused on what we call *models* rather than formally developed theories. This is, in part, because laws—general and abstract explanatory principles—play a minor role. Traditional views have held that events are explained by showing they are to be expected as the result of natural laws.[1] More recently philosophers have moved away from an emphasis on laws, understood formally. This is sometimes described as a *semantic view of theories*. Models are introduced as intermediate structures, somewhere between laws in a classical sense and data. Bas van Fraassen explains it this way:

> To present a theory is to specify a family of structures, its *models*; and secondly, to specify certain parts of those models (the *empirical substructures*) as candidates for the direct representation of observable phenomena. (1980, p. 64)

Models here are abstract structures; they can be thought of as the (nomologically) possible worlds of which a theory is true. Laws are yet more abstract, leaving specific parameters and parameter values open. In Ronald Giere's (1988) treatment of classical mechanics, the explanatory models are constructed from Newton's laws, though not by any simple derivation; rather, they are developed by fixing some parameter values and introducing a number of simplifying assumptions and approximations. As a consequence the models do not precisely fit any real situation, but to the extent that they approximate real situations, they can be used to understand them.

Some philosophers who have focused on evolutionary biology have defended a similar perspective. In particular, John Beatty (1980, 1981), Elizabeth Lloyd (1984, 1986, 1988, 1989), and Paul Thompson (1983, 1989) have argued that evolutionary biology, population genetics, and evolutionary ecology are better represented in terms of a semantic view. Insofar as we emphasize models, we are not unsympathetic to semantic views. In the cases we have described, however, we find even less of a role for laws. Construing these models as constructed from more abstract laws by fixing parameter values, even with additional simplifying assumptions and approximations, if practicable at all, would be little more than an idle exercise, and one accomplished after the fact. The explanatory task begins and ends with the models.[2] Models in this sense are akin to blueprints; they are partial and abstract representations of the causal mechanisms. A scientist constructs a model by envisaging various parts, which can be partially described (sometimes as a consequence of experimental probes, sometimes by appealing to analogies with other more familiar mechanisms), and conceptualizing how such parts will interact with each other. The explanatory power of a model stems from its ability to show how some phenomenon or range of phenomena would be the consequence of the proposed mechanism.

By developing a model of mechanistic explanation, our goal has not been to construct a general theory of explanation, or even of mechanistic explanation. Neither has it been to offer any general conception of theories. We do claim that mechanistic models are often the vehicles of explanation in the biological and psychological sciences and that they often constitute what scientists count as theories. But our focus has been on the construction of causal explanations; that is, on the development of mechanistic models. In a suitably broad sense this can be understood as a concern with the development of theories. As we suggested in Chapter 1, it is in the dynamics of theories that we can hope to find an account of scientific discovery.[3] We have so far focused on the kinematics of change. We have described a number of decisions, or choice-points, that define the direction of development. These (illustrated in Figure 8.7) are less chronological than logical choice-points.

Two of the choice-points particularly involve decomposition and localization. The first, which occupied us in Part II, is the identification of a system as the locus of control and the determination of the functional properties of components within the system. Identifying the cell as the locus of control for respiration resolved itself into the question whether respiration was carried out only in specialized loci (for example, the lungs or the blood), or in the tissues as well. As we explained in Chapters 3 and 4, resolving this issue was partly an experimental question and partly a question shaped by more general metaphysical and theoretical commitments. The experimental questions focused on the capacities of specific tissues in isolation. The more general commitments ranged from a commitment to mechanistic explanations—and especially to an antivitalism assumed by most of the major contenders—to more specific commitments shaped by thermodynamic and chemical theories of the time.

The second step, which occupied us in Part III, involves the analysis of the activities of the controlling system into component functions. As with identifying the locus of control, this decomposition must be combined with a localization that resolves the system into appropriate units and assigns relevant functions to them. We have seen that a number of outcomes are possible, ranging from simple localization of a function in a component of the whole system, to models emphasizing interaction of components. If the components are thought to operate primarily on their own, we have component systems; if the parts are so linked that the activities of one are regularly altered by the activities of others, we have integrated systems. Once again the specific outcome depends only partly on experimental results; it also incorporates an array of more general commitments. Successful mechanistic explanations require convergence of the decomposition into component functions with successful localization of these components within the system. Developing such convergence often relies on a wide array of experimental and theoretical constraints. These can range from broadly correlational constraints, as in Chapter 6, to more restrictive physical constraints, as in Chapters 7 and 8. At both the point where one chooses to segregate a system from its environment and attribute an activity to it, and the point where one tries to decompose the function and localize it in components of the system, it is possible to seek out alternatives to mechanistic explanation. While we have discussed some of these alternatives, our primary focus has been on the development of mechanistic models.

While we have partially succeeded if the choice-points we have identified provide an accurate account of the kinematics of theory development, our ultimate goal is to go further and develop a dynamic analysis of theory development. This requires us to identify the constraints that play the role of forces governing theory development. The kinematics constitute the phenomena that a dynamic model should explain.

2. FOUR CONSTRAINTS ON DEVELOPMENT

The cases we have discussed can provide a rough taxonomy of the array of constraints that affect the development of mechanistic models; all can be encapsulated in a dynamic model of theory development. This model is speculative and provisional. The constraints we identify, though, are not only consistent with, but are suggested by, the more descriptive, kinematic picture. They fall into four basic groups, summarized in Figure 10.1. The first are *psychological* constraints, including the heuristic strategies employed. If we are right, localization and decomposition are among them. The second are what we will describe as *phenomenological* regularities. These consist of, primarily, information concerning the behavioral capacities of the system, including the effects of perturbations. The third are *operational* constraints, which encompass the experimental procedures and models that limit what can be asked about the behavior of components. The fourth are *physical* constraints, which limit the range of allowed component functions. In what follows we will say a bit more about these four constraints and their role in a more general model of theory change.

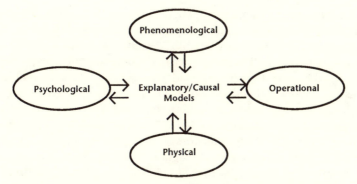

Figure 10.1. Four Constraints on Explanatory Models. There are at least four constraints influencing the development of causal and explanatory models. This provides at most a rough indication of the general categories included and exemplified in one or another of the cases discussed in previous chapters.

Psychological Constraints

Psychological constraints are largely heuristic, in the sense sketched in Chapter 1. Decomposition and localization are heuristics that we have identified as applicable to the development of mechanistic explanations of complex systems. They serve, at least, to simplify the explanatory problem. They are domain specific, and while they are not unconditionally conducive to realistic models, they sometimes lead to realistic causal explanations. One reason decomposition and localization may fail is that the assumptions they impose—that the system is decomposable or nearly decomposable—may be false. Some systems are decomposable, and some are nearly so; others are less so, though assuming that they *are* decomposable is one way to discover that they are not. Insofar as these heuristics are domain specific, they are also liable to error. We saw in Chapter 9 that there is at least some reason to think that decomposition and localization do *not* reveal the actual organization of some systems, neurophysiological, psychological, and genetic. Decomposition and localization are relatively restrictive and so quite powerful. That is, they significantly limit the search space of possible explanatory models to ones with reasonably simple organization. Accordingly, they provide a relatively strong guide on the development of explanatory models. When they fail, it is necessary to expand the range of explanatory models under consideration.

The critical point is that these heuristics, like others, reflect cognitive strategies through which humans approach a complex problem. They are necessary in part because human information-processing capacities are limited, and the available information is even more limited. The limitations of both human and machine intelligence were some of the first motivations for introducing heuristics into AI modeling. In these models there is assumed to be a well-defined space of possible solutions, and the task is to search this space to find the equation or law that solves the problem. If we assume a well-defined and adequate search space, then all discovery requires is a means to search through the space of allowed solutions. Accordingly, BACON searches a limited space of mathematical relations, and DENDRAL assumes a determinate set of organic molecules. Even in a search of a well-defined space, though, the search time might exceed the information-processing capacities of humans or machines, and heuristics are used to focus the search on the most promising parts of that space.

In the cases we have considered there is no antecedently specifiable set of mechanisms, and hence no well-defined space to search for possible mechanisms. Indeed, we do not even know whether the space we are searching contains a solution at all. Decomposition and localization guide the search for an adequate model within that space; that is, they guide and constrain the construction of explanatory models. Another, distinct, limitation is critical here: the data vastly underspecifies the nature of the mechanism. Thus, the researcher must develop strategies for finding plausible models given the data. Decomposition and localization simultaneously provide such a strategy. They not only lead researchers to begin with the most manageable models, but also with ones that have high probative value. By assuming decomposition and seeking a direct localization, scientists can be led to models that are easy to manage and which often guide the search for additional factors that must be explained by a more complete model. Hypothesizing that nature is nearly decomposable, and developing linear models, reduce the cognitive demands and can lead to models with areas in which organization is more complex and in which component integrity is undermined. Thus, even failures in decomposition and localization may be guides to fruitful models of integrated component systems.

We emphasized that the need for heuristics stems from the limitations of human information-processing. The actual heuristics employed by a scientist, however, are not determined by these limitations alone. Intellectual and cultural backgrounds also shape the heuristics actually utilized. In particular, the restriction to mechanistic models is culturally conditioned. This affects both the experimental methods that are applied and the theoretical applications that accompany them. What counts as a mechanistic model, moreover, can and does change. Cartesian mechanism excluded action at a distance, and Newton certainly never contemplated the array of mechanical forces that are now commonplace.

Phenomenological Constraints

In more traditional, empiricist-inspired accounts of scientific explanation, what scientists are thought to explain is a body of data. In the simplest construal of the deductive-nomological model, the data is taken to be a class of observations. The inference from data to theory may be inductive or abductive; either existing or novel observations serve as data against which theories are measured (cf. Bechtel 1988). It is now commonplace to note that "data" is theory laden. Since Hanson (1958) and Kuhn (1962) drew our attention to the phenomenon, philosophers and historians have accepted that data assumes a significance only in light of the theoretical assumptions brought to it. At the very least the theoretical background provides the conceptual apparatus in terms of which the data is understood and described. Yet, it is still often assumed that the goal is to explain the data; that is, to derive the observations from some law or set of laws.

Just as we do not deny the existence and usefulness of laws, we do not deny the existence and usefulness of data. Data in a relatively unproblematic sense is often useful in guiding the search for models and is regularly employed in evaluating proposed explanations. However, data are not the *explananda* of scientific models. We observe data, but we explain phenomena. Phenomena are not data in any useful sense. Phenomena are *repeatable* features of the world which reveal themselves in a variety of experimental arrangements. If one arrangement leads us to think that there is a body of phenomena that cannot be revealed within other paradigms, we suspect that we are confronted with an artifact. We rely on data to identify phenomena to be explained and again to evaluate models. As James Bogen and James Woodward explain in an insightful essay, explanatory models do not predict or explain the data we observe, although that data provides relevant information about the phenomena:

> Data, which play the role of evidence for the existence of phenomena, for the most part can be straightforwardly observed. However, data typically cannot be predicted or systematically explained by theory. By contrast, well-developed scientific theories do predict and explain facts about phenomena. Phenomena are detected through the use of data. (1988, pp. 305–6)

Except in special cases, such as when we try to explain why an experiment did not give us the results expected, we do not, and cannot, explain data. We may have what we think is a reliable means of detecting some phenomenon, and in the process we may rely on data, but we may not be able to explain how we got the data or why it is a reliable indicator of the phenomenon. To explain the data would require knowing, for example, how the instruments are used in the procurement of data work. But we relied on microscopes before we knew how they worked, and thus before we could explain the data they generated (Hacking 1983). Whereas data are "idiosyncratic to particular experimental contexts," phenomena "have stable, repeatable characteristics which will be detectable by means of a variety of different procedures" (Bogen and Woodward 1988, p. 317). Data are ephemeral. Phenomena are robust.

Cartwright's following defense of "phenomenological laws" is a defense of phenomena in the sense we intend:

> Phenomenological laws describe what happens. . . . For the physicist, unlike the philosopher, the distinction between theoretical and phenomenological has nothing to do with what is observable and what is unobservable. Instead the terms separate laws which are fundamental and explanatory from those that merely describe. (1983, p. 2)

Pre-Copernican astronomers spoke of "saving the phenomena," meaning both the more regular motions of heavenly bodies and the less regular occurrences such as eclipses, occlusions, and retrogression. *Phenomenological* thermodynamics includes within its domain the regularities that govern the behavior of gases, such as the Boyle-Charles Law and much more. Though some researchers would offer an empiricist rendering of these laws (Duhem 1906; van Fraassen 1980), reducing them to the data points from which they were derived, data in the sense of the products of individual observations and experiments are not what scientists set out to explain. Hacking illustrates the difference between a datum and a phenomenon:

> A phenomenon is *noteworthy*. A phenomenon is *discernible*. A phenomenon is commonly an event or process of a certain type that occurs regularly under definite circumstances. The word can also denote a unique event that we single out as particularly important. When we know the regularity exhibited in a phenomenon we express it in a law-like generalization. (1983, p. 221)

There is feedback between phenomena and explanatory models. Because something must be noteworthy or detectable as a regularity to be a phenomenon, what counts as a phenomenon to be explained depends on what explanatory models are available and alive. In acknowledging this, however, we do not intend to endorse the constructivism found in Hacking's comment that "experiment is the creation of phenomena" (ibid., p. 229). Ours is an ecumenical realism. Minimally, a phenomenon is a regularity in a system's behavior. The fact that yeast cells ferment, and that they do so at a particular rate under specified conditions, is a phenomenon that calls for explanation. So is the fact that a trait is heritable within a particular environment. The fact that tissues can respire in the absence of blood is another phenomenon that must be explained by a theory of respiration. Again, the fact that hippocampal lesions affect behavior of rats on complex mazes more than on simple mazes is a phenomenon that should be explained by a model of spatial-problem solving. All of these are phenomena to be explained. They are real, and so, we think, are their causes.

Phenomena are an important constraint on the attempt to develop an explanation. An explanation is erroneous if it fails to account for them. They speak with a voice, though, that is subject to interpretation. In recent debates over the architecture of cognition, some theorists claim that any explanation of the phenomena surrounding cognition must invoke symbolic structures and rules. Fodor and Pylyshyn (1988), for example, contend that the productivity and systematicity exhibited by cognitive phenomena *require* a system with a certain architecture. While the characteristics of the phenomenon to be explained may be suggestive as to possible structures of the explanatory model, they are not sufficient to necessitate a particular form to the model. As we observed earlier, Justus Liebig attempted to formulate the structure of a general model of animal chemistry on the basis of the phenomenon of animal nutrition: animals consume foodstuff containing more complex chemical structures than are found in their waste product. Liebig concluded that all the reactions in the animal body accordingly had to be catabolic. Subsequently, Bernard demonstrated the occurrence of glycogenesis and argued for the more general phenomenon of synthetic reactions in the animal. This meant not only that the explanatory model needed revision, but that metabolism needed to be understood in different terms. Metabolism was not simple decomposition of foodstuff. Likewise, classical genetics construed its phenomena in terms of the transmission of phenotypic traits. One consequence of Beadle and Tatum's work was the recognition that the phenotype, and the phenomena of genetics, must themselves be understood in biochemical terms. Phenomena, as well as the explanatory models, change. This is what we have called reconstituting the phenomena (see Chapter 8).

Operational Constraints

We remarked in Chapter 2 that in complex systems, and particularly in self-organizing systems, the nature of the components and their contributions cannot generally be inferred from normal behavior of the system alone. Interaction between components in a smoothly functioning system makes any simple inference from the phenomena to the mechanism problematic. What is required are independent means for assessing and understanding the behavior of components. Often a great deal of resourcefulness is needed in the development and utilization of experimental procedures to study a system and its components.

The problems in determining the components and their contributions are often practical and not theoretical in any meaningful sense. It is often not possible to isolate components physically, and it would be uninformative if we did, for they would no longer be doing what they do in the normal system. We would, for example, gain little information about the function of the liver by simply cutting it out of the body, unless we have developed means for interacting with it that help to reveal its natural operation. What we require are techniques that show what the parts of a system are and how they function within the system.

Often what are required are appropriate instruments and experimental techniques. To measure the respiration of organisms, Lavoisier and La-Place had to develop a calorimeter. To measure the respiration promoted by Atmungsferment, Warburg had to develop a special flask. To determine the role of the hippocampus in spatial memory, O'Keefe and Nadel had to carefully lesion the hippocampi in experimental rats. Not only must instruments and techniques be developed, but they must be used in a manner that is truly probative. It was no small flaw in Flourens's technique that he lacked aseptic methods, and it is not small virtue of Pasteur that he did not. At other times what is required is a useful model system. In some cases a model system might be a natural system that is particularly useful because it is easy to work with or thought to provide a good example of the phenomenon. The use of insect models in determining the locus of respiration, and of *Drosophila* as a useful genetic model, both played critical roles in the development of mechanistic models. And sometimes the model systems are artificial systems that are thought to simulate real systems and provide demonstrations of how they might work.

The use of instruments, techniques, and model systems are all prone to the production of artifacts. For experimental scientists this is a continuous worry. Part of the reason this concern looms large is that the use of instruments, techniques, or model systems cannot generally be justified by appeal to well-developed and tested theories about how they work. They are generally widely employed long before researchers understand why they work as they do. Rather, they are often justified by more indirect means. For example, the data they give may seem so clean that researchers are convinced that it must reflect the phenomena under study. In other cases the picture they provide of phenomena is corroborated by evidence produced by other techniques. Finally, sometimes the fact that they provide evidence that makes sense in terms of emerging models is seen as vindicating their use (Bechtel 1990).

Although we have not focused in detail on the introduction of instruments and research techniques, or on the factors used to evaluate potential model systems, we have discussed two general strategies. One approach depends on the disruption of system behavior in order to find what deficit from the normal behavior of the system results. We have termed this an inhibitory method. In the simplest form, this approach involves the observation of deficits in the presence of damage to the system, as when we observe linguistic deficits with trauma to the left frontal lobe. In most instances, however, the case is more indirect. O'Keefe and Nadel inferred the function of the hippocampus, in part, from the behavior of animals suffering hippocampal lesions, assuming that they would compensate for the damage. They sought what Jackson called *positive* symptoms rather than simple deficits. The requirement that we be able to isolate intermediaries noted in Chapters 7 and 8, is a relatively strong operational constraint. Abnormal excretions, as in alkaptoneuria, provide strong evidence about metabolic intermediaries in intact systems. Meyerhof's use of flouride to inhibit muscle glycolysis, with a resulting buildup of hexosediphosphate, was evidence of the latter's importance as a metabolic intermediary.

The second experimental approach we discussed follows the opposite course, providing an unusually strong stimulus in order to elicit evidence of what a component of a system does. We have termed this an *excitatory* method. Artificial stimulation of the cortex, as in the classic work of Fritsch and Hitzig (1870), provides relatively direct information about centers of control in the nervous system. The injection of metabolic intermediaries is similar. An important part of the positive case for the role of phosphates in glycolysis depends on Embden's demonstration that an artificial increase of hexosediphosphate leads to an increase in lactic acid.

As we have noted, the issue of how the results should be interpreted arises in both inhibitory and excitatory studies. The fact that a deficit occurs from an inhibitory study does not prove what the inhibited part contributes to the normally operating system. The deficit may in fact be the result of a variety of interactions in the system that occur in an atypical fashion due to the interruption. This is particularly true in integrated systems. Similarly, the increased performance, or the failure to generate an increased output, from an excitation does not categorically point to the (non)contribution of the excited component. For example, the addition of intermediates to the fermentation system did not necessarily result in an increase in alcohol production, because this activity required the integration of different components, not just the one being excited.

We have focused on only two strategies for the development of mechanistic models. While we do not intend this to be a complete list of research tools used in such investigations, we do want to draw attention to the fact that what can be experimentally accessed is as important as what might be theoretically recognized. Science is, in large measure, a practical art.

Physical Constraints

Bogen and Woodward insist that scientific explanations exhibit "detailed dependencies":

> It is not a satisfactory explanation of an outcome merely to assert that it is due to some general mechanism, where the details of the mechanism are left unspecified. Instead, a satisfactory systematic explanation must show how the features of the explanandum-phenomenon systematically depend upon the factors invoked in the explanans of that explanation. (1988, p. 323)

When we are dealing with models developed through the heuristics of decomposition and localization, we embrace strong assumptions about how the system is physically realized. It is an important aspect of the cases discussed in Part III that evidence for the physical realizability of the component functions was required. This entails our having evidence from the lower level for the mechanisms posited in the decomposition: Wernicke defended his explanation of the aphasias in part on the basis of Meynart's work in neuroanatomy. Researchers investigating fermentation required that each step in a proposed model involve an independently known chemical reaction. As we noted, it was one of the primary virtues of Neuberg and Kerb's model that it met this constraint. Garrod's work, as we have also mentioned, faced the same constraint.

The insistence on physical realizability is suggestive of reduction. In any useful sense of the term, we do accept that decomposition and localization constitute a reductionistic strategy. Mechanistic explanations *do* explain the activities of a system in terms of the behavior of component parts. Decomposition and localization are not reductionistic, however, in the sense that is current in the philosophical literature. Traditional literature has focused on laws stated as universally quantified statements and on whether the laws of one theory can be derived from those of another. In addition to the derivation of one set of laws from another, philosophical models of reduction require that *bridge laws* connect the terms of one theory with those of the other. There has, for example, been a great deal of controversy within philosophy over whether Mendelian genetics has actually been reduced to molecular genetics, or whether psychology might thus reduce to neuroscience. Much of this controversy focuses on whether we can establish appropriate bridge laws linking molecular and Mendelian concepts (cf. Hull 1974; Kitcher 1984; Wimsatt 1976). It appears that Mendelian concepts fail to map smoothly onto molecular mechanisms, so that the same Mendelian mechanism is subserved by different molecular mechanisms, and the same molecular mechanism may serve different Mendelian functions in different contexts. Philip Kitcher declares flatly,

The main difficulty in trying to axiomatize classical genetics is to decide what body of statements one is attempting to axiomatize. The history of genetics makes it clear that Morgan, Muller, Sturtevant, Beadle, McClintock, and others have made important contributions to genetic theory. But the statements occurring in the writings of these workers seem to be far too specific to serve as parts of a general theory. (1984, p. 351)

We have no doubt that the cases we have discussed, including the work in biochemical genetics, will fail to fit the philosophical mold that has been cast for reduction. The investigators we have examined were not concerned with developing general laws, and when they did turn to the lower level it was not in order to derive antecedently developed laws at the higher level. Rather, they turned to the lower level to create models of mechanisms that would explain specific processes observed at a higher level. Thus, such scientists are subject to the very objection Kitcher advances. Because we regard the philosophical model as not much more than a philosophical construct, having little to do with scientific practice, this leaves us unconcerned. Whether it counts as reduction or not, we think that the history of genetics makes it transparently clear that the work of Morgan, Muller, Sturtevant, Beadle, McClintock, and others *was* part of a general theory of development and heredity, and that this general theory *was* intended to explain development and heredity in biochemical terms.

3. Conclusion: Looking Forward

Our efforts have been directed at reaching a realistic account of the development of science. If we have been successful, then we have correctly characterized the kinematics of one sort of research program—one directed toward developing mechanistic explanations in the face of complex natural systems. We have, more speculatively, indicated four factors, or types of factors, that feature in the dynamics of theories. More is still to be done in developing a dynamic model and in elaborating the component forces.

The focus on decomposition and localization takes us in the direction of models that span more than one level. As we have seen, decomposition and localization are fallible, as are all heuristics; but even when they fail, they may serve as probative tools for facilitating discovery. The emphasis on the lower-level mechanisms in explaining processes identified and conceptualized at higher levels is what many scientists think of as reductionistic. In any case, lower-level constraints are essential in developing and elaborating explanatory models. Absent such constraints, either for lack of appropriate research tools or because of a failure to pursue experimental inquiries at lower levels, we are left with speculative theories. We have seen not only how a failure to incorporate lower-level constraints is problematic, but also how such constraints have been brought to bear in elaborating and developing interlevel mechanistic explanations. We have offered a qualitative model of theory development designed to account for the processes by which scientists develop mechanistic models of phenomena of interest. This account is meant to be psychologically and historically realistic. The choices we have identified are intended as points in the development of research programs at which decisions shape the course of research and the formulation of models.

The result may not suit a general philosophical temperament. We offer no universal procedures, no single method for science. We see no invariant pattern. Scientific disciplines evolve under differing constraints with differing histories. Their actual development, as historians have long seen, cannot be understood apart from the changing historical context. This does not mean the process is ungoverned or arbitrary. As Peter Galison (1990) suggests, the development of scientific theories is not so much plastic as it is "immensely constrained." If we are right, the problem is one of reaching a solution that simultaneously satisfies a complex array of changing constraints. An account of theory development that takes such constraints seriously requires a measured historicism. Such is also a realistic historicism.

Notes

CHAPTER 1
COGNITIVE STRATEGIES AND SCIENTIFIC DISCOVERY

1. In retrospect, the more prominent reasons for repudiating any "logic of discovery" are unconvincing and rest on a distinction between discovery and justification that will not sustain critical scrutiny. Reichenbach is typical in claiming that the "act of discovery escapes logical analysis" and grounding this on the view that "there are no logical rules in terms of which a 'discovery machine' could be constructed" (1951, p. 231). As we will note, there are programs that do model scientific discovery, in the sense that they employ procedures capable of inducing general laws from limited data; however, if a "logical analysis" in the sense Reichenbach intends requires an effective procedure for discovery, there would be no "logic of discovery." Neither would there be a "logic of justification" in this sense, since that would require an effective procedure for determining whether an arbitrary claim is a consequence of a theory. There is, of course, no such procedure even for moderately complex formal systems. Producing a justification is itself a process of discovery.

2. For a recent treatment emphasizing this point see Longino (1990), where the importance of differing background assumptions is carefully demonstrated as applied to sexual differences and their supposed causal etiology.

3. There was considerable research into maternal, cytoplasmic, inheritance both before and after the rise of American genetics at the turn of the century. American geneticists refined (or defined) the problems of genetics in such a way that the relevant problems were restricted to those of transmission. This is an interesting episode, told well by Jan Sapp (1987).

4. The classic study is Meehl (1954); for more recent studies, see Dawes, Faust, and Meehl (1989), or Faust and Ziskin (1988).

5. Tversky and Kahneman likewise expose the importance of their heuristics by illustrating deviations from what is "normatively appropriate." This clearly does not entail that, for example, we use Bayesian rules when we are right and appeal to representativeness or salience only when we err.

6. Lindley Darden accurately notes (in personal correspondence) that this is not true of chess programs. Some of the best systems have traded on the computer's ability and patience for wide and deep search. It is also true that those that emphasize wide and deep search are remarkably unrevealing about human expertise. For recent discussions, see Charness (1989).

7. A little knowledge goes a long way. Feigenbaum (1989) explains that, for twenty carbon atoms, mass spectrometry was able to "prune" the search space to approximately a million plausible candidates. Once given information about the number of methyls and their connections to carbon and nitrogen atoms, the number of candidates was reduced to one.

8. In later versions the program is more flexible, incorporating multiple independent variables and multiple levels of description.

CHAPTER 2
COMPLEX SYSTEMS AND MECHANISTIC EXPLANATIONS

1. Liebig's contemporaries often expressed their doubts about the adequacy of his approach in unmistakable terms. Berzelius, for one, derided Liebig's work as "physiological chemistry . . . created at the writing table" (quoted in Fruton 1972, p. 97).

2. If Liebig is to be faulted, it is for undervaluing empirical data, not for advancing a speculative model.

3. Strictly, any classification of organizational properties must be relativized to a specific decomposition and to some specific set of systemic properties (cf. Kauffman 1976). A system may, for example, be aggregative for one set of systemic properties, but not for another; likewise, it may be aggregative for a set of properties under one decomposition, but fail to be aggregative for the same set of properties under another. Analogous points can be made for more complex forms of organization. When relativization is pertinent, we shall introduce it explicitly.

4. Wimsatt has told us that he clearly would *not* count some of the cases as nonaggregative that we *do* see as such. To some extent we see the discrepancies between the ways we and Wimsatt deal with interactions as resulting from differences in our projects. Wimsatt seems primarily interested in marking off "emergent" systems, while we are concerned with exploring variations in organizational structure. We do not have the space to elaborate on these differences in any detail here.

CHAPTER 3
IDENTIFYING THE LOCUS OF CONTROL

1. Not always, of course. Longer term effects—such as the threat of radioactive contamination, or even the development of heart disease—are ones we do not readily bring to bear in decision making, presumably because these effects are distant. As Tversky and Kahneman (1957) put it, such distant effects lack "salience."

2. Robert Brandon makes the point clearly in the case of evolutionary biology, distinguishing the "selective environment" from the "ecological" and "external" environments; he shows that for many purposes the important variable is the selective environment (1990, ch. 2).

3. We will subsequently see that both associationism and functionalism had substantial impact on the development of ideas about the operation of the nervous system. This connection, however, lies outside our current concern.

4. Watson was a student of James Angell and H. H. Donaldson at the University of Chicago, completing his thesis in 1903 on the maturation of the white rat and its influence on learning ability. For a useful overview, see Boakes (1984, ch. 6), where both the emphasis on comparative psychology and many of the influences on Watson are put into perspective.

5. According to Watson's view, the basic patterns of association, or conditioned learning, are determined by the central nervous system. This is a problem for physiology rather than psychology.

6. In the early editions of the *Origin* this phrase was largely applied to degenerative changes resulting in the reduction of organs. In later editions and other works, such as *The Variation of Plants and Animals Under Domestication* (1868), it was applied more broadly.

7. In this respect variation is not random for Darwin. But this in no way compromises his view that variation is random *with respect to adaptation*: the mechanisms that induce variation, for example, might make a mutation more common than its reverse mutation, without any implication that one is better adapted than the other.

8. They are available in any variety of current texts. For a very brief summary, see Richardson (1984, pp. 411ff.).

9. Natural languages, Chomsky told us, are potentially infinite, and they appear to incorporate rewrite rules that are context sensitive. This would require relatively powerful generative systems.

10. We do not intend to suggest that the debates between preformationists and epigenesists have taken the same form repeatedly. We are actually faced with two evolving families of views, with the positions and the issues shifting over time. For two treatments we find especially useful, see Maienschein (1986a), and Roe (1981).

11. Cope and Hyatt independently developed recapitulationist accounts of evolution in 1866, the same year as the publication of Haeckel's *General Morphology of Organisms*.

12. Both Cope and Hyatt saw a pattern in the process of evolution. They thought it was essential for an evolutionary mechanism to explain the pattern, and correspondingly they can be seen to be committed to orthogenesis; however, this does not make their view goal-directed, and there is little reason to think they saw the pattern as progressive (see Richardson and Kane 1988; also Kane and Richardson 1985).

13. See Gould (1977) for a historical overview. In Chapter 9 we will discuss a variant on the controversy.

14. Our discussion here draws heavily on the historical analyses of Culotta (1970a, b) and Keilin (1966).

15. Beginning here is a matter of convenience. We do not mean to suggest that Lavoisier's theory can or should be understood apart from the historical context in which it arose. For a general discussion of Lavoisier, see Holmes (1985). For a useful discussion on the contrasting views of Joseph Priestley, see McEvoy (1988); McEvoy and McGuire (1975).

16. The chemical techniques that showed the oxygen binding powers of hemoglobin and the absorption spectrum of oxyhemoglobin were developed by Hoppe (1862) and Stokes (1864) only after Bernard's experiments were reported.

17. Bernard was also skeptical of Liebig's results on the grounds that they were in vitro experiments and might therefore reflect artifacts produced in the laboratory rather than the process as it occurs in living organisms.

18. Like Bernard, Ludwig's commitment to mechanism was uncompromising.

He was a member of the group composed also of Helmholtz, du Bois Reymond, and Brücke, who in the late 1840s called for the development of a totally mechanistic approach to physiology, modeled as much as possible on physics (see Cranefield 1957).

19. Like Bernard, Ludwig advanced evidence that oxygen did not pass from the blood into the tissue based on the fact that vacuum pumps did not seem capable of effectively deoxygenating blood, showing that it was difficult to decouple oxygen from blood.

CHAPTER 4
DIRECT LOCALIZATION

1. We do not mean to suggest that this is a surprising correlation. It is exactly what IQ tests are constructed to predict. That they do so is nonetheless an empirical fact. We also do not mean to suggest, and do not in fact believe, that differences in IQ explain differences in scholastic performance.

2. As Cooter (1984) observes, this hardly meant that the Societies or the BAAS were lacking phrenologists in their ranks, or that phrenology went undiscussed in their meetings. However, phrenology's status was ambivalent at best. For example, when the BAAS met in Edinburgh in 1834, Adam Sedgwick exhorted its members "to confine their researches to dead matter, without entering into any speculations on the relations to intellectual beings" (cited in Cooter 1984).

3. Gall also made fundamental and important contributions to neuroanatomy and comparative anatomy. Though these were important in supporting the credibility of phrenology, as Young emphasizes, they were at best loosely connected with Gall's phrenological views.

4. The doctrine hardly began with Descartes or his contemporaries. Galen held the more radical view that the substance of the brain was responsible for the activities of mind, and Vesalius had left Descartes's view on a reasonably firm empirical footing by the early part of the sixteenth century.

5. John D. Davies (1955) offers a useful discussion of phrenology as a social movement in the United States, and David de Guistino (1975) incorporates a similar discussion of its relation to Victorian social thought. Harrington (1987) provides a description of the importance of phrenology to issues in psychology during the last half of the nineteenth century.

6. In the United States phrenology was subjected to repeated abuses from the pulpit for its affinities with atheism, materialism, and determinism. In Europe, as in America, the views attained a great deal of popular attraction and enlisted support from such scientific luminaries as Alfred Russel Wallace and Benjamin Silliman, even if the views received less than unqualified support from the intellectual and medical communities.

7. This proved to be a source of conflict between Gall and Spurzheim, along with the latter's deviant taxonomy of primary faculties.

8. The most common objection to this method is that sinus cavities will defeat any useful cranioscopic measures.

9. In a shift that foreshadowed some of the most innovative features of Hughlings Jackson's views (which we will discuss in Chapter 9), Spurzheim also argued

that functional disorders are not the effect of primitive faculties, but "the effect of predominance of powers."

10. The term *enzyme* literally means "in yeast." The word was introduced in the context of a controversy over whether chemical agents (enzymes) or whole living organisms were required for physiological reactions such as fermentation. For further discussion, see Chapter 7.

11. When Wieland did receive the Nobel Prize in Chemistry in 1927, it was not for his work on biological oxidation, but for his research on articulating the structure of bile acids.

12. Wieland thus exemplifies a common pattern: it is often outsiders who are responsible for major reconceptualizations in a particular field. An outsider, moreover, often faces problems in gaining acceptance within a research tradition, as did Wieland.

13. Holmes (1986) shows how biochemists in the first two decades of this century developed and pursued a research program directed toward identifying linear sequences of reactions in intermediary metabolism, modeled on the b-oxidation of fatty acids.

14. Initially Warburg (1911) took this result to confirm Overton's (1899) theory that narcotics affect cell activity by affecting the physical state of the lipids in the cell. Subsequently, Warburg found that narcotics combined more rapidly with the solid particles.

15. Warburg's oxidation work was confined to the amino acids, but Meyerhof (1924) provided evidence that the charcoal preparations could oxidize hexosephosphates, although not simple glucose.

16. His use of carbon monoxide also differentiated it from another iron component of the cell that had then been recently discovered, Keilin's cytochrome. Cytochrome does not interact with carbon monoxide and is present in far greater quantities than the reactive iron. Keilin's discovery in 1925 that cytochromes were reversibly oxidized and reduced in cellular respiration posed another competitor to Warburg's proposal. Warburg initially sought to disprove any role of cytochromes, until it was finally established that his Atmungsferment was, in fact, Keilin's "cytochrome oxidase," or what later was known as "cytochrome a_3," an auto-oxidizable cytochrome that is sensitive to carbon monoxide (see Keilin 1966; Warburg 1949).

17. It was for this work that Warburg was awarded the Nobel Prize for Physiology of Medicine in 1931 (see his Nobel Lecture [1931] for an overview of this work).

18. He seldom hired biochemists to work in his laboratory, preferring to hire pure technicians to assist in his work (see Krebs 1981).

19. As noted above, Wieland offered a different explanation of the efficacy of hydrogen cyanide, claiming that it had a greater affinity than oxygen for the active enzyme. He also pointed to evidence from Dixon and Elliot (1929) indicating that hydrogen cyanide only inhibited about 50% of respiration in pulp preparations. Alt provided Warburg with an answer to this problem, showing that the partial inhibition was due to the experimental circumstances that allowed some of the cyanide to be lost. When proper precautions were taken, inhibition reached Warburg's predicted 95% level.

20. We do not claim that these are the only factors shaping the formulation of the problem. There was no hope at Gall's time of an empirically based functional anatomy of the brain, and a structural decomposition of the cell was beyond the reach of Wieland and Warburg.

CHAPTER 5
THE REJECTION OF MECHANISM

1. For an enlightening discussion of the debate between Haller and Wolff, emphasizing the philosophical and religious concerns that informed their debates in the eighteenth century, see Shirley Roe's (1981) discussion of the topic.

2. Between 1822 and 1828 Cuvier read at least four positive reports on Flourens's experimental neurophysiology to the Académie des Sciences, and, with Cuvier's support, Flourens was elected to the Académie in December 1828. Flourens also was Cuvier's *suppléant* at the Collége de France, eventually succeeding Cuvier both at the Collége and as secretary of the Académie.

3. The craniological commitments get virtually no mention and seem to play no significant role in the debates during the period.

4. These are not unreasonable complaints, in retrospect. With Flourens's methods, however skillfully executed, there must have been considerable damage to other structures. Indeed, as Young (1970, pp. 60–61) points out, Flourens's extirpations ignored cortical structure. Furthermore, the sceptic conditions would inevitably have led to generalized destruction of brain tissue (for further discussion, see Tizard 1959).

5. Albury (1977) provided a careful examination of Bichat's experimental procedure. Bernard often contrasted Magendie and Bichat in terms of their differing attitudes toward experimentation. Albury showed that parallels can be found in Bichat's research to several of Magendie's major experimental works. According to Albury, the difference between the two lay their explanatory perspectives: Bichat sought to attribute a function to a particular agent; Magendie felt the goal was not to find an agent for a function, but to explain functions in terms of functions.

6. There is irony in this development, because prior to the 1830s fermentation had not even been associated with living organisms. Willis (1684) and Stahl (1697/1748), for example, both explained fermentation as the result of the violent internal motion of particles within the fermenting substance, a motion that resulted in the decomposition of the substance into simpler components.

7. They were not the first to offer evidence for yeast being a living organism, but they were the first to attract serious attention. Earlier, Desmazieres (1826) had characterized yeast globules as living organisms, and Colin (1825) had shown that fermentation could occur without oxygen. The attractiveness of the mechanical conception of fermentation as a natural oxidation, however, kept interest in the possibility that yeast might be a living organism at a minimum.

8. As we shall see in Chapter 7, by the early years of the twentieth century alcoholic fermentation and lactic acid fermentation were recognized to be virtually identical. The primary difference between the two lies in the final steps. In the former, one pyruvic acid is decarboxylated and then reduced to alcohol; in the other it is directly reduced to lactic acid. This was not known until the 1930s.

PART III
INTRODUCTION

1. Other research groups have failed to find the same genetic markers in other populations. This is taken to suggest that there is more than one gene that predisposes patients to the mental disorder.

2. It has since turned out that the initial results of the Amish study are indeed spurious correlations. New clinical data has undermined the result (see Kelsoe et al. 1989).

CHAPTER 6
COMPLEX LOCALIZATION

1. For related work, see Gluck and Bower (1988); Gluck and Thompson (1987); Rescorla (1984).

2. The disturbance of speech under localized lesions had already been described by Bouillaud, as well as by Gall (1810–1819, 4:83).

3. The cases are examined more fully in Richardson (1980 and 1986a).

4. Wernicke is silent on concepts pertaining to abstract objects, but it seems reasonable to assume that his account of perceptible properties could be extrapolated from his view of concepts of concrete objects.

5. It is here we have true mimicry, repression, and modification of responses—or, conscious control of behavior. It may be of interest to compare this with Hughlings Jackson, whose work is discussed in Chapter 9, section 2.

6. Simple correlations of behavioral deficits with underlying neural pathologies are, of course, rare. As Wernicke saw, this is fundamentally a consequence of the fact that damage to the nervous system is rarely restricted to single neural structures. Vascular damage, for example, does not generally correspond to functional neuroanatomical structures, even if it happens to in some cases. Gunshot wounds are even less discriminating.

7. It is also not inconsistent with Lenneberg's claim, which is developmental. That is, Lenneberg claims that recovery of function at earlier ages is a matter of relearning. The claim is not without its problems, but they are not germane here. Even if there is recovery of function, and the "recovered function" is realized differently than was the original function, decomposability is preserved. Equipotentiality imposes a very strong condition to the effect that other parts of the system must perform *the same function in the same way*. Similarity is not enough.

8. Ferrier's "Scientific Phrenology," in *The Functions of the Brain* (1876), is a striking development within this tradition, though it did not influence Wernicke as far as we know.

9. Further clinical observation led Wernicke to complicate this simple scheme considerably.

10. EEG records electrical activity from the scalp and is a general measure of brain activity. Statistical techniques can be used to identify specific cortical areas generating EEG activity.

11. These correlations are, of course, also consistent with other hypotheses, such as David S. Olton's claim that the hippocampus is responsible for working

memory (cf. Olton, Becker, and Handelmann 1979). As we have already noted, it is not necessary for us to resolve the current disputes over the proper functions of the hippocampus.

12. There are also afferents from the diagonal band of Broca and various portions of the brain stem.

13. It is important to recognize that O'Keefe and Nadel's characterization *cannot* be adequately understood as a simple, two-level theory. Some of the information used—for example, the monitoring of "place" detectors—bridges several levels of organization.

14. The parallel to Wernicke should be clear: discrete systems will manifest their characteristic symptoms only in unusual circumstances.

CHAPTER 7
INTEGRATED MECHANISMS

1. See Chapter 2 above for the story of Hora and Tempus. The point is that if a system is assembled from its parts it is more efficient if the component parts form stable subassemblies and do not disintegrate before being incorporated into the larger system. This yields organization that is not only decomposable, but hierarchical.

2. One of the most accessible illustrations of the importance of homology in evolutionary theorizing is Stephen Gould's title trilogy in *The Panda's Thumb* (1980).

3. This is not to imply that there is another discrete unit—an executive—exercising control. Pattee says that the demands imposed on a higher-level unit would define and determine the behavior of components; we say, instead, that the explanation of system behavior depends on organization rather than constituent function.

4. The cases discussed in Chapter 6 also have a linear organization, with information feeding in only one direction.

5. Researchers were keenly aware that they had to provide independent evidence of the occurrence of the intermediate reactions they posited, and of the existence in living organisms of the enzymes that catalyzed them.

6. The general term *ferment* was used to refer to the agent responsible for the chemical action, whether that agent was a chemical substance or a living cell.

7. The living protoplasm theory was thus a vitalistic theory, though one which resisted any dualism of substances (cf. Bechtel 1984b).

8. This lactic acid fermentation was the object of Pasteur's first research on fermentation (for discussion, see Chapter 5).

9. The use of the same vocabulary to describe the overall process and its constituent processes is interesting. The constituent processes are in fact quite different in character from the overall process, and as biochemistry matured, different vocabularies came to be used, with reference to intermediaries "fermenting" gradually disappearing.

10. The early-twentieth-century biochemists here recapitulated the assumption made by Liebig in the nineteenth century. Liebig described animal nutrition

in the 1840s as involving only catabolic reactions in which energy-rich foodstuff was broken down to release energy.

11. As with any of the claims that substances could or could not be fermented, the actual results were mixed. Different results had to do with subtle differences in technique. For details of these studies, see Harden (1932) or Florkin (1975).

12. The methyl was thought to be derived from the methylglyoxal that figured as an intermediary in step 2. Neuberg also engaged in a number of inhibitory studies to show that his postulated intermediaries were actually produced in biological fermentations by fixing and removing them from living cells. For example, he employed sodium sulfate to prevent the reduction of acetylaldehyde and show its occurrence in fermenting material (Neuberg and Reinfurth 1920).

13. Meyerhof admitted that this was only an ad hoc hypothesis tailored to account for the physiological data and that he lacked a chemical interpretation of the notion of a stabilization stage.

14. One interesting issue, which we shall not pursue here, is why the linear representation of the pathway is still preferred to the cyclic representation. The choice of one representational form is surely suggestive of some guiding heuristic.

15. Developing the explanation at this level does not obviate the need to consider lower levels. As work proceeded on the physical constitution of enzymes, investigation turned to the mechanisms of catalytic reactions.

16. Other researchers, including Dorothy Needham, showed a similar willingness to speculate about the underlying chemical processes on the basis of metabolic needs of the cell, even in the absence of any direct knowledge of the physical basis of the mechanisms.

CHAPTER 8
RECONSTITUTING THE PHENOMENA

1. We use "rediscovery" lightly. For a discussion that is both entertaining and enlightening, see Robert Olby (1985, chapter 6; and also the appendix to chapter 5).

2. For discussions of the implications of Mendelism oriented toward philosophical issues, see especially Darden and Maull (1977), and Darden (1980 and 1991) and Kitcher (1984).

3. Allen (1978) tells us that the first clear evidence of linkage, noting both complete and partial linkage between distinct traits, was reported by Bateson in 1906.

4. For an interesting discussion of Ephrussi, see Jan Sapp's (1987); for a perspective somewhat different from ours, see Burian, Gayon, and Zallen (1988 and forthcoming).

5. We have relied in what follows on a detailed study by Olby (1974). Those interested in the case or related research should consult his work.

6. This fits comfortably with the view, common at the time, that recessive characters are the result of inactive or malfunctional genes; that is, that in the absence of dominant alleles, recessive characters will be manifested. It would then be natural to find recessive genes expressed in close marriages, as that would be more likely to result in dual recessives. Garrod does not use the term *gene*, which Lin-

dley Darden points out (in personal correspondence) was not introduced until 1909. See note 7 for a related point.

7. This is particularly obvious in light of Garrod's inattention to deviations from anticipated Mendelian ratios in the inheritance of alkaptonuria. Such deviations constituted the central phenomena for the Morgan school, and it is both important and interesting that Garrod did *not* raise the questions one might expect were he working within a Morganist tradition.

8. It can also be seen in the earlier work of Lucien Cuénot (1903). Cuénot notes that some offspring of albino mice are not albinos, a result that is impossible if albinism is a simple Mendelian recessive. Once pigmentation is seen as the product of several genes, though, this result is intelligible. Albinism can result from recessive mutants at more than one locus. Interbreeding can then restore the pathways in offspring.

9. Morgan reported radium-induced mutations as early as 1914, and Muller demonstrated the feasibility of inducing mutations with X-rays in 1927.

10. It should be remembered that it was not thought at this time that genes were nucleotide sequences, but instead were thought to be proteins or nucleoproteins. It was therefore open whether genes merely determined enzyme specificities, or whether they were themselves enzymes.

11. It is a mistake to think of this as *eliminativist* in the way philosophers use the expression. No one contends—or even suggests—that there are not flies with vermilion eyes. That verges on the absurd. It is merely that eye color per se ceases to be a Mendelian trait and to assume the central role it once did.

CHAPTER 9
"EMERGENT" PHENOMENA IN INTERCONNECTED NETWORKS

1. As we have seen, such equipotentiality provides significant comfort for antimechanists, as in Flourens's ablation studies (see Chapter 4).

2. For useful discussions from quite different perspectives, see (Smith 1982b), Star (1983), and Young (1970). We are especially indebted to the latter source.

3. For a useful discussion of Spencer's psychology and its relation to his evolutionary speculations, see Richards (1987, ch. 6).

4. As we mention in Chapter 6, Wernicke also depended on associationistic principles.

5. It was typical of nineteenth-century evolutionary thinking to hold that progressive change involved the addition of structures later in development and to emphasize the parallels between ontogeny and phylogeny (cf. Gould 1977).

6. Spencer describes the development of mental life as an evolutionary progression from reflex and instinct through memory to reason and will. These capacities differed only in their complexity.

7. We must be careful with Jackson's use of the word *representation*. It does not signify the existence of an autonomous module that has internal models of what it represents. If it did, we could analyze it independently and explain the system by carrying out a typical decomposition. According to Jackson's view, however, all that seems to be involved is that the higher-level structure is designed to respond

appropriately to the lower-level units, and these in turn are responsive to the activities of the higher-level centers.

8. For an interesting, if rather eccentric, discussion of Jackson, see Head (1926, 1:330ff.).

9. For extended analyses of these connectionist approaches, see McClelland, Rumelhart, and the PDP Research Group (1986); and for a slightly different approach, see Grossberg (1982 and 1988). For an exposition of connectionist systems and discussion of their significance, see Bechtel and Abrahamsen (1991) and Churchland (1989).

10. It is also necessary to add some mechanism for decreasing the strength of the connections. This can be done simply, by imposing a spontaneous decay in the connection strength, or in other, more elaborate, ways.

11. This simulation was performed using the software provided by McClelland and Rumelhart (1988). The results are reported in more detail by Bechtel and Abrahamsen (1991).

12. The most successful symbolic models have employed numeric parameters on rules, locating both existing exemplars and new examples in a multidimensional similarity space. See Anderson (1983) and Medin and Schaffer (1978).

13. Back-propagation is a generalization of the delta rule discussed above, both of which modify connection strengths in a direction that will reduce the overall error.

14. A cycle consists of a randomly ordered presentation of the 100 training patterns.

15. After one training set Hinton's network was correct on all four remaining propositions; in another it was correct on three out of four. The difference is due to the stochastic feature of the training process: if the network begins with different random weights, or the training instances are presented in different orders, a different set of weights can result.

16. For a more general treatment of Kauffman's work, see Burian and Richardson (1991).

17. Relative fitnesses are stable.

18. The lac operon in *E. coli* is evidently a canalyzing function. In the presence of allolactose, the operator is free.

19. It is certainly no coincidence that successful development of connectionist models, either for cognition or for genetics, has coincided with the availability of powerful, high-speed computer systems. Without them, projecting the consequences of constellations of changes would certainly be overwhelming.

CHAPTER 10
CONSTRUCTING CAUSAL EXPLANATIONS

1. Of course, this is given a linguistic turn. The account starts not with the phenomenon to be explained, but with a description of it. One explains the phenomenon by deriving the description of it from laws and descriptions of initial conditions.

2. This is not to deny that there are cases in which laws do play central or important roles. For example, in developing models of biochemical processes,

laws of thermodynamics and kinematics are applied to specify the effects of the model system. We question the hegemony of laws in explanation, not their existence.

3. A "theory" in this broad and general sense is not a formal set of laws, but a matrix of changing models bound together partly by history and partly by a shared commitment within a community to a general framework and method.

References

Albury, W. R. 1977. Experiment and Explanation in the Physiology of Bichat and Magendie. In W. Coleman and C. Lisnoges, eds., *Studies in the History of Biology*. Baltimore: Johns Hopkins University Press.

Alkon, D. L. 1989a. *Memory Traces in the Brain*. Cambridge: Cambridge University Press.

———. 1989b. Memory Storage and Neural Systems. *Scientific American* 261:42–50.

Allen, G. E. 1978. *Thomas Hunt Morgan: The Man and His Science*. Princeton: Princeton University Press.

Anderson, J. R. 1983. *The Architecture of Cognition*. Cambridge: Harvard University Press.

Angell, J. R. 1907. The Provence of Functional Psychology. *Psychological Review* 14:61–91.

Anonymous. 1839. Das enträthselte Geheimniss der geistigen Gährung. *Justus Liebig's Annalen der Chemie* 29:100–104.

Appel, T. A. 1987. *The Cuvier-Geoffroy Debate: French Biology in the Decades Before Darwin*. New York: Oxford University Press.

Appert, N. 1810. *Le Livre de tous les Ménages ou l' Art de Conserver, pendant plusieurs Années toutes les Substances Animales et Végétablés*.

Bach, A. 1897. Du Rôle des Peroxydes dans les Phénomènes d' Oxydation Lente. *Comptes Rendus Hebdomadaires des Séances de l' Academie des Sciences, Paris*. 124:951.

———. 1913. Über den Mechanismus der Oxydationsvorgänge. *Berichte der deutschen chemischen Gesellschaft* 46:3864.

Barnes, B. 1974. *Scientific Knowledge and Sociological Theory*. London: Routledge & Kegan Paul.

Barnes, D. M. 1987. Defect in Alzheimer's is on Chromosome 21. *Science* 235:846–47.

Bateson, W. 1909. *Mendel's Principles of Heredity*. Cambridge: Cambridge University Press.

Bateson, W., and E. R. Saunders. 1902. *Reports to the Evolution Committee of the Royal Society, London* 1:1–160.

Battelli, F., and L. Stern. 1911. Die Oxydation der Bersteinsäure durch Tiergewebe. *Biochemische Zeitschrift* 30:172–94.

Beadle, G. W. 1945. Biochemical Genetics. *Chemical Reviews* 37:15–96.

———. 1959. Genes and Chemical Reactions in Neurospora. *Science* 129:1715–19.

Beadle, G. W., and B. Ephrussi. 1936. The Differentiation of Eye Pigments in *Drosophila* as Studied by Transplantation. *Genetics* 21:225–47.

———. 1937. Development of Eye Colors in *Drosophila:* Diffusible Substances and Their Interactions. *Genetics* 22:76–86.

Beadle, G. W., and E. L. Tatum. 1941a. Experimental Control of Development and Differentiation: Genetic Control of Developmental Reactions. *American Naturalist* 75:107–16.

———. 1941b. Genetic Control of Biochemical Reactions in *Neurospora*. *Proceedings of the National Academy of Science* 27:499–506.

Beatty, J. 1980. Optimal-Design Models and the Strategy of Model Building in Evolutionary Biology. *Philosophy of Science* 47:532–61.

———. 1981. What's Wrong with the Received View of Evolutionary Theory. In P. D. Asquith and R. N. Giere, eds., *PSA 1980: Proceedings of the 1980 Biennial Meeting of the Philosophy of Science Association*, 2:397–426. East Lansing: Philosophy of Science Association.

Bechtel, W. 1982. Two Common Errors in Explaining Biological and Psychological Phenomena. *Philosophy of Science* 49:265–92.

———. 1984a. The Evolution of Our Understanding of the Cell: A Study in the Dynamics of Scientific Progress. *Studies in the History and Philosophy of Science* 15:309–56.

———. 1984b. Reconceptualization and Interfield Connections: The Discovery of the Link Between Vitamins and Co-enzymes. *Philosophy of Science* 51:265–92.

———. 1986a. Biochemistry: A Cross-Disciplinary Endeavor that Discovered a Distinctive Domain. In W. Bechtel, ed., *Integrating Scientific Disciplines*, 77–100. Dordrecht: Martinus Nijhoff.

———. 1986b. Building Interlevel Pathways: The Discovery of the Embden-Meyerhof Pathway and the Phosphate Cycle. In J. Dorn and P. Weingartner, eds., *Foundations of Biology*, 65–97. Vienna: Holder-Pichlert-Tempsky.

———. 1988. *Philosophy of Science: An Overview for Cognitive Science*. Hillsdale, N.J.: Erlbaum.

———. 1989. An Evolutionary Perspective on the Re-Emergence of Cell Biology. In K. Halweg and C. A. Hooker, eds., *Issues in Evolutionary Epistemology*, 433–57. Albany, N.Y.: SUNY Press.

———. 1990. Scientific Evidence: Creating and Evaluating Experimental Instruments and Research Techniques. In A. Fine, M. Forbes, and L. Wessels, eds., *PSA 1990: Proceedings of the 1990 Biennial Meeting of the Philosophy of Science Association*, 1:559–72. East Lansing: Philosophy of Science Association.

Bechtel, W., and A. A. Abrahamsen. 1991. *Connectionism and the Mind: An Introduction to Parallel Distributed Processing*. Oxford: Basil Blackwell.

Bechtel, W., and R. C. Richardson. 1992. Emergent Phenomena and Complex Systems. In A. Beckermann, H. Flohr, and J. Kim, eds., *Emergence or Reduction? Essays on the Prospects of Nonreductive Physicalism*, 257–88. Berlin and New York: Walter de Gruyter Verlag.

Bernard, C. 1855. Sur le Mécanisme de la Formation du Sucre dans le Foie. *Comptes Rendus Hebdomadaires des Séances de l' Académie des Sciences* 41:461–69. Translated in Gabriel and Fogel (1955), 93–97.

———. 1856. Recherches Expérimentales sur la Température Animal. *Comptes Rendus Hebdomadaires des Séances de l' Académie des Sciences* 43:332.

———. 1859. *Lescons sur les Propriétés Physiologiques et les Alterations Pathologiques des Liquides de l' Organisme*. Paris: Baillière.

———. 1865. *An Introduction to the Study of Experimental Medicine*. Reprint. New York: Dover, 1957.

———. 1877a. *Leçons sur le Diabète et la Glycongenèse Animale*. Paris: Baillière.

———. 1877b. Critique Expérimentale sur le Mécanisme de la Formation du Sucre dans le Foie. *Comptes Rendus Hebdomadaire des Séances de l' Académie des Sciences, Paris* 85:519–25.

Berthelot, M. 1859. Remarques sur la Fermentation Alcoolique de la Levure de Bière. *Comptes Rendus Hebdomadaires des Séances de l' Académie des Sciences, Paris* 48:691–92.

———. 1860. *Chimie Organique Fondée sur la Synthèse*. Paris: Mallet-Bachelier.

———. 1873. Remarques sur un Point Historique Relatif a la Chaleur Animale. *Comptes Rendus Hebdomadaires des Séances de l'Académie des Sciences, Paris* 77:1065.

———. 1889. Sur la Chaleur Animale-Chaleur dégagée par l'Action de l'Oxygéne sur le Sang. *Comptes Rendus Hebdomadaires des Séances de l' Académie des Sciences, Paris* 109:776–81.

Berzelius, J. J. 1836. Einige Ideen über bei der Bildung organischer Verbindungen in der lebenden Naturwirksame, aber bisher nicht bemerkte Kraft. *Jahres-Berkcht über die Fortschritte der Chemie* 15:237–45.

———. 1837. Über eine bei der Bildung Organischer Verbindungen wahrscheinlich Wirksame, bis jetzt wenig bemerkt Kraft. *Lehrbuch der Chemie*. Translated by F. Woehler.

———. 1839. Weingährung. *Jahres-Bericht über die Fortschritte der Chemie* 18:400–03.

Bichat, X. 1801/1822. *Anatomie Générale Appliquée à la Physiologie et à la Médecine*. Paris: Brosson, Gabon et Cie. Partially reprinted in Hall (1951), 68–72.

———. 1805. *Recherches Physiologiques sur la Vie et la Mort*. 3d ed. Paris: Marchant.

Boakes, R. 1984. *From Darwin to Behaviorism*. Cambridge: Cambridge University Press.

Bogen, J., and J. Woodward. 1988. Saving the Phenomena. *Philosophical Review* 97:303–52.

Bowler, P. 1988. *The Non-Darwinian Revolution: Reinterpreting a Historical Myth*. Baltimore: Johns Hopkins University Press.

Bradie, M. 1986. Assessing Evolutionary Epistemology. *Philosophy and Biology* 1:401–59.

Braine, M.D.S. 1978. On the Relation between the Natural Logic of Reasoning and Standard Logic. *Psychological Review* 85:1–21.

Brande, W. T. 1806. Chemical Experiments on Guaiacum. *Philosophical Transactions*, 89–98.

Brandon, R. 1978. Adaptation and Evolutionary Theory. *Studies in History and Philosophy of Science* 9:181–206.

———. 1990. *Adaptation and Environment*. Princeton: Princeton University Press.

Broca, P. 1861a. Perte de la Parole. *Bulletins de la Société Anthropologie* 2:235–38.

Broca, P. 1861b. Remarques sur le Siêge de la Faculté Suivies d' une Observation d' Aphémie. *Bulletins de la Société Anatomique de Paris* 6:343–57. Partially translated by Mollie D. Boring in R. J. Herrnstein and E. G. Boring, eds., *A Source Book in the History of Psychology*, 223–29. Cambridge: Harvard University Press, 1965.

Brown, J. W. 1980. Brain Structure and Language Production: A Dynamic View. In D. Caplan, ed., *Biological Studies of Mental Processes*, 287–300. Cambridge: M.I.T. Press.

Buchanan, G. G., G. L. Sutherland, and E. A. Feigenbaum. 1969. Heuristic DENDRAL: A Program for Generating Explanatory Processes in Organic Chemistry. In B. Meltzer and D. Michie, eds., *Machine Intelligence 4*. New York: Elsevier.

Buchner, E. 1897. Alkoholische Gährung ohne Hefezellen Vorläufige Mittheilung. *Berichte der deutschen chemischen Gesellschaft* 30:117–24. Partially translated in Gabriel and Fogel (1955), 27–30.

Buchner, E., and J. Meisenheimer. 1904. Die chemische Vorgänge bei der alkoholischen Gärung. *Berichte der deutschen chemischen Gesellschaft* 37:417–28.

Buchner, H. 1889. Lecture to the Munich Society of Physicians, November 21, 1888. *Münich Med Wochschr*, 36.

Burian, R., J. Gayon, and D. Zallen. 1988. The Singular Fate of Genetics in the History of French Biology, 1900–1940. *Journal of the History of Biology* 21:357–402.

———. Forthcoming. Ephrussi and the Synthesis of Genetics and Embryology. In S. Gilbert, ed., *A Conceptual History of Embryology*. New York: Plenum.

Burian, R., and R. C. Richardson. 1991. Form and Order in Evolutionary Biology: Stuart Kauffman's Transformation of Theoretical Biology. In A. Fine, M. Forbes, and L. Wessels, eds., *PSA 1990: Proceedings of the 1990 Biennial Meeting of the Philosophy of Science Association*, 2:267–87. East Lansing: Philosophy of Science Association.

Cagniard-Latour, C. 1838. Memoire sur la Fermentation Vineuse. *Annales de Chimie*, 2d ser., 68:206–23.

Carlson, N. 1966. *The Gene: A Critical History*. Philadelphia: W. B. Saunders.

———. 1981. *Physiology of Behavior*. Boston: Allyn and Bacon.

Cartwright, N. 1983. *How the Laws of Physics Lie*. Oxford: Clarendon Press.

Charness, N. 1989. Expertise in Chess and Bridge. In D. Klahr and K. Tovsky, eds., *Complex Information Processing*, 183–208. Hillsdale, N.J.: Lawrence Erlbaum.

Chase, W. G., and H. Simon. 1973a. Perception in Chess. *Cognitive Psychology* 4:55–81. As reprinted in Simon (1979), 386–403.

———. 1973b. The Mind's Eye in Chess. In W. G. Chase, ed., *Visual Information Processing*, 215–81. San Diego, Calif.: Academic Press. As reprinted in Simon (1979), 404–27.

Cheng, P. W., and K. J. Holyoak. 1985. Pragmatic Reasoning Schemas. *Cognitive Psychology* 17:391–416.

———. 1989. On the Natural Selection of Reasoning Theories. *Cognition* 33:285–313.

Chi, M.T.H., R. Glaser, and M. J. Farr. 1988. *The Nature of Expertise.* Hillsdale, N.J.: Lawrence Erlbaum.

Chomsky, N. 1957. *Syntactic Structures.* The Hague: Mouton & Co.

———. 1965. *Aspects of a Theory of Syntax.* Cambridge: M.I.T. Press.

———. 1968. *Language and Mind.* Enl. ed. New York: Harcourt Brace Jovanovich.

———. 1975. *Reflections on Language.* New York: Pantheon Books.

———. 1980. *Rules and Representations.* New York: Columbia University Press.

Churchland, P. M. 1989. *A Neurocomputational Perspective: The Nature of Mind and the Structure of Science.* Cambridge: M.I.T. Press/Bradford Books.

Churchland, P. S. 1986. *Neurophilosophy.* Cambridge: M.I.T. Press/Bradford Books.

Colin, J. J. 1825. Mémoire sur la Fermentation du Sucre. *Annales de Chimie* 28:128–42.

Combe, G. 1835. *A System of Phrenology.* Boston: Marsh, Capen, and Lyon.

Connant, J. B. 1948. Pasteur's study of fermentation. In J. B. Connant, ed., *Harvard Case Studies in Experimental Science,* 437–85. Cambridge: Harvard University Press.

Cooter, R. 1976. Phrenology. *History of Science* 14:211–34.

———. 1980. Deploying Pseudoscience: Then and Now. In M. P. Hanen, M. J. Osler, and R. G. Weyant, eds., *Science, Pseudoscience and Society.* Waterloo, Ont.: Wilfrid Laurier.

———. 1984. *The Cultural Meaning of Popular Science: Phrenology and the Organization of Consent in Nineteenth-Century Britain.* Cambridge: Cambridge University Press.

Cope, E. D. 1887. *The Origin of the Fittest.* New York: Appleton.

Correns, C.F.J.E. 1900. Mendel's Regel über das Verhalten der Nachkommenschaft der Rassenbastarde. *Berichte der deutschen botanischen Gesellschaft* 18.

———. 1903. Über die dominierenden Merkmale der Bastarde. *Berichte der deutschen botanischen Gesellschaft* 21:133–47.

Cranefield, P. F. 1957. The organic physics of 1847 and the biophysics of today. *Journal of the History of Medicine* 12:407–23.

Cuénot, L. 1903. L' Hérédité de la Pigmentation chez les Souris. *Archs Zool. exp. gén.* 1:33–41.

Culotta, C. A. 1970a. Tissue Oxidation and Theoretical Physiology: Bernard, Ludwig, and Pflüger. *Bulletin of the History of Medicine* 44:129–38.

———. 1970b. On the Color of the Blood from Lavoisier to Hoppe-Seyler, 1777–1864: A Theoretical Dilemma. *Episteme* 4:219–33.

Cummins, R. 1983. *The Nature of Psychological Explanation.* Cambridge: M.I.T. Press/Bradford Books.

Cuvier, G. 1812. *Recherches sur les Assemens Fossiles de Quadrupèdes, où l' on Rétablit les Caractères de Plusieurs Espèces d' Animaux que les Révolutions du Globe Paroissent Avoir Détruites.* 4 vols. Paris.

Dakin, H. D. 1912. *Oxidations and Reductions in the Animal Body.* London: Longmans, Green, and Company.

———. 1922. *Oxidations and Reductions in the Animal Body.* 2d ed. London: Longmans, Green, and Company.

Darden, L. 1980. Theory Construction in Genetics. In T. Nickles, ed., *Scientific Discovery: Case Studies*, 151–70. Dordrecht: D. Reidel.

———. 1986. Relations among Fields in the Evolutionary Synthesis. In W. Bechtel, ed., *Integrating Scientific Disciplines*, 113–23. Dordrecht: Martinus Nijhoff.

———. 1991. *Theory Change in Science: Strategies from Mendelian Genetics*. Oxford: Oxford University Press.

Darden, L., and N. Maull. 1977. Interfield Theories. *Philosophy of Science* 43:44–64.

Darden, L., and R. Rada. 1988. Hypothesis Formation Using Part-Whole Interrelations. In D. Helman, ed., *Analogical Reasoning*. Dordrecht: D. Reidel.

Darwin, C. 1859. *On the Origin of Species*. 1st ed. facsimile. Cambridge: Harvard University Press, 1964.

———. 1868. *The Variation of Plants and Animals Under Domestication*. 2 vols. New York.

———. 1871. *The Descent of Man, and Selection in Relation to Sex*. Reprinted with an introduction by J. T. Bonner and R. M. May. Princeton: Princeton University Press, 1981.

———. 1872. *The Origin of Species*. 6th ed. New York.

Davies, J. D. 1955. *Phrenology, Fad, and Science: A Nineteenth Century Crusade*. New Haven: Yale University Press.

Davy, E. 1820. On Some Combinations of Platinum. *Philosophical Transactions* 110:108–25.

Davy, H. 1817. Some New Experiments and Observations on the Combustion of Gaseous Mixtures, with an Account of a Method of Preserving a Continued Light in Mixtures of Inflammable Gases and Air without Flame. *Philosophical Transactions* 107:77–85.

Davy, J. 1815. Versuche über das Blut. *Deutsches Archiv für Physiologie Meckel* 1:109–16.

Dawes, R. M., D. Faust, and P. E. Meehl. 1989. Clinical Versus Actuarial Judgment. *Science* 243:1668–74.

Dennett, D. C. 1978. *Brainstorms*. Cambridge: M.I.T. Press/Bradford Books.

Desmazieres. 1826. *Ann Sci Naturelles* 10: 42–67.

Dijksterhuis, E. J. 1950. *The Mechanization of the World Picture*. Princeton: Princeton University Press.

Dixon, M., and K.A.C. Elliot. 1929. The Effect of Cyanide on the Respiration of Animal Tissues. *Biochemical Journal* 23:812–29.

du Bois-Reymond, E. 1859. Über die Angeblick saure Reaktion des Muskelfleisches. *Monatsbericht die Akadamie*, 288–324. Reprinted in *Gesammelte Abhandlungen*, vol. 2. Freiberg.

Duclaux, E. 1896. *Pasteur: Histoire d'un Esprit*. Sceaux: Charaire.

Duda, R. O., and E. H. Shortliffe. 1983. Expert Systems Research. *Science* 220:261–68.

Duhem, P. 1906. *The Aim and Structure of Physical Theory*. Reprint. New York: Atheneum, 1954.

Egeland, J. A., et al. 1987. Bipolar Affective Disorders Linked to DNA Markers on Chromosome 11. *Nature* 325:783–87.

Eggert, G. E. 1977. *Wernicke's Works on Aphasia: A Source Book and Review.* The Hague: Mouton & Co.

Eggleton, P., and G. P. Eggleton. 1927. The Physiological Significance of Phosphagen. *Biochemical Journal* 63:155–61.

Einbeck, H. 1914. Über das Vorkommen der Fumarsäure im freschen Fleische. *Hoppe-Seylers Zeitschrift für physiologische Chemie* 90:303–7.

Ellis, Daniel. 1807. *An Inquiry into the Changes Induced on Atmospheric Air, by the Germination of Seeds, the Vegetation of Plants, and the Respiration of Animals.* Edinburgh.

Embden, G., H. J. Deuticke, and G. Kraft. 1933. Über die intermediaren Vorgänge bei der Glykolyse in der Muskulatur. *Klinische Wochenschrift* 12:213–15. Reprinted in Kalckar (1969), 67–72.

Embden, G., W. Griesbach, and E. Schmitz. 1914. Über Milchsäuerebildung und Phosphorsäurebildung im Muskelpreßsaft. *Zeitschrift für physiologische Chemie* 93:1–45.

Embden, G., F. Kalberlah, and H. Engel. 1912. Über Milchsäurebildung im Muskelpreßsaft. *Biochemische Zeitschrift* 45:45–62.

Engler, C., and W. Wild. 1897. Ueber die sogenannte Aktivierung des Sauerstoffs und ber Superoxydbildung. *Berichte der deutschen chemischen Gesellschaft* 30:1669.

Engler, C., and Wissberg. 1904. *Kritische Studien ber die Vorgänge der Autoxydation.* Braunschweig: Vieweg und Sohn.

Ephrussi, B. 1942. Chemistry of 'Eye Color Hormones' of *Drosophila. Quarterly Review of Biology* 17:327–38.

Farley, J., and D. L. Alkon. 1985. Cellular mechanisms of Learning, Memory, and Information Storage. *Annual Review of Psychology* 36:419–94.

Faust, D. 1984. *The Limits of Scientific Reasoning.* Minneapolis: University of Minnesota Press.

Faust, D., and J. Ziskin. 1988. The Expert Witness in Psychology and Psychiatry. *Science* 241:31–35.

Feigenbaum, E. A. 1989. What Hath Simon Wrought? In D. Klahr and K. Kotovsky, eds., *Complex Information Processing*, 165–82. Hillsdale, N.J.: Lawrence Erlbaum.

Feigenbaum, E. A., and J. Feldman, eds. 1963. *Computers and Thought.* New York: McGraw Hill.

Fernbach, A., and M. Schoen. 1913. *Comptes Rendus Hebdomadaires des Séances de l' Académie des Sciences, Paris* 157:1478.

Ferrier, D. 1876. *The Functions of the Brain.* London.

Fisher, R. A. 1930. *The Genetical Theory of Natural Selection.* Oxford: Clarendon Press.

Fiske, C. H., and Y. Subbarow. 1929. Phosphorus Compounds of Muscle and Liver. *Science* 70:381–82. Reprinted in Kalckar (1969), 54–56.

Fletcher, W. M., and F. G. Hopkins. 1907. Lactic Acid in Amphibian Muscle. *Journal of Physiology* 35: 247–309. Partially reprinted in Kalckar (1969): 318–25.

Florkin, M. 1975. *Comprehensive Biochemistry.* Vol. 31, *A History of Biochemistry.* Part III, *History of the Identification of the Sources of Free Energy in Organisms.* Amsterdam: Elsevier.

Flourens, J.P.M. 1824. On the Functions of the Brain. *Recherches Expérimentales sur les Propriétés et les Fonctions du Système Nerveux dans les Animaux Vertébrés*, 236–41.

———. 1846. *Phrenology Examined*. Philadelphia: Hogan and Thompson. Republished in D. N. Robinson, ed., *Significant Contributions to the History of Psychology, Series E: Physiological Psychology*. Washington, D.C.: University Publications of America, 1978.

Fodor, J. A. 1975. *The Language of Thought*. New York: Thomas Y. Crowell.

———. 1983. *Modularity of Mind*. Cambridge: M.I.T. Press/Bradford Books.

Fodor, J. A., and Z. W. Pylyshyn. 1988. Connectionism and Cognitive Architecture: A Critical Analysis. *Cognition* 28:3–71.

Foster, M. 1901/1970. *Lectures in History and Physiology During the Sixteenth, Seventeenth and Eighteenth Centuries*. New York: Dover.

Fowler, O. S. 1848. *Practical Phrenology*. New York: Nafis & Cornish.

Fraassen, B. van. 1980. *The Scientific Image*. Oxford: Clarendon Press.

Fritsch, G., and E. Hitzig. 1870. Über die elektrische Erregvarkeit des Grosshirns, *Archiv für Anatomie, Physiologie und Wissenschaftliche Medizin*. English translation in G. von Bonin, *Some Papers on the Cerebral Cortex*, 73–96. Springfield, Ill.: Charles C. Thomas, 1960.

Fruton, J. 1972. *Molecules and Life: Historical Essays on the Interplay of Chemistry and Biology*. New York: Wiley Interscience.

Gabriel, M. L., and S. Fogel. 1955. *Great Experiments in Biology*. Englewood Cliffs, N.J.: Prentice Hall.

Galison, P. 1989. "Multiple Constraints, Simultaneous Solutions." In A. Fine and J. Leplin, eds., *PSA 1988: Proceedings of the 1988 Biennial Meeting of the Philosophy of Science Association*. 2:157–63. East Lansing: Philosophy of Science Association.

Gall, F. J., and J. C. Spurzheim. 1810–1819. *On the Functions of the Brain and Each of Its Parts: with Observations on the Possibility of Determining the Instincts, Propensities, and Talents, or the Moral and Intellectual Dispositions of Man and Animals, by the Configuration of the Brain and the Head*. 6 vols. Translated into English by Winslow Lewis Jr. Boston: Marsh, Capen, and Lyon, 1835.

Garrod, A. 1899. A Contribution to the Study of Alkaptonuria. *Medico-Chirurgical Transactions* 82:367–94.

———. 1902. The Incidence of Alkaptonuria: A Study in Chemical Individuality. *The Lancet* 2:1616–20.

———. 1908. Inborn Errors of Metabolism (The Croonian Lectures). *The Lancet* (4 July): 4, 1–7; (11 July): 73–79; (18 July): 142–48; (25 July): 214–20.

Gay-Lussac, J. L. 1810. Extrait d' un Mémoire sur la Fermentation. *Annales de Chimie* 76:245–59.

Gerhardt, C. 1856. *Traite de Chimie Organique*. Paris: Firmin Didot.

Geschwind, N. 1964. The Development of the Brain and the Evolution of Language. Reprinted in Geschwind (1974), 86–104.

———. 1965. Disconnexion Syndromes in Animals and Man. Reprinted in Geschwind (1974), 105–236.

———. 1967. Wernicke's Contribution to the Study of Aphasia. Reprinted in Geschwind (1974), 284–98.

————. 1974. *Selected Papers on Language and the Brain*. Dordrecht: D. Reidel.

————. 1980. Some Comments on the Neurology of Language. In D. Caplan, ed., *Biological Studies of Mental Processes*, 301–19. Cambridge: M.I.T. Press.

Ghiselin, M. 1974. A Radical Solution to the Species Problem. *Systematic Zoology* 23:536–44.

Giere, R. 1979. *Understanding Scientific Reasoning*. New York: Holt, Reinhart, & Winston.

————. 1988. *Explaining Science*. Chicago: University of Chicago Press.

Gilmartin, K., and H. Simon. 1973. A Simulation for Memory of Chess Positions. *Cognitive Psychology* 5:29–46. As reprinted in Simon (1979), 373–85.

Glas, E. 1979. *Chemistry and Physiology in Their Historical and Philosophical Relations*. Netherlands: Delft University Press.

Gluck, M., and G. Bower. 1988. From Conditioning to Category Learning, an Adaptive Network Model. *Journal of Experimental Psychology, General* 117:227–47.

Gluck, M., and R. F. Thompson. 1987. Modeling the Neural Substrates of Associative Learning and Memory: A Computational Approach. *Psychological Review* 94:176–91.

Goldgaber, D., et al. 1987. Characterization and Chromosomal Localization of a cDNA Encoding Brain Amyloid of Alzheimer's Disease. *Science* 235:877–80.

Goldman, A. I. 1986. *Epistemology and Cognition*. Cambridge: Harvard University Press.

Goodfield, G. J. 1960. *The Growth of Scientific Physiology*. London: Hutchinson.

Gould, S. J. 1977. *Ontogeny and Phylogeny*. Cambridge: Harvard University Press.

————. 1980. *The Panda's Thumb*. New York: W. W. Norton & Co.

————. 1981. *The Mismeasure of Man*. New York: W. W. Norton & Co.

Gould, S. J., and R. C. Lewontin. 1979. The Spandrels of San Marco and the Panglossian Paradigm: A Critique of the Adaptationist Programme. *Proceedings of the Royal Society of London* B205:581–98.

Gregory, R. L. 1961. The Brain as an Engineering Problem. In W. H. Thorpe and O. L. Zangwill, eds., *Current Problems in Animal Behavior*, 307–30. Cambridge: Cambridge University Press.

————. 1968. Models and the Localization of Function in the Central Nervous System. Reprinted in C. R. Evans and A.D.J. Robertson, eds., *Key Papers: Cybernetics*, 91–102. London: Butterworths.

————. 1981. *Mind in Science*. Cambridge: Cambridge University Press.

Grene, M. 1974. *The Knower and the Known*. Berkeley and Los Angeles: University of California Press.

Grobstein, C. 1973. Hierarchical Order and Neogenesis. In H. H. Pattee, ed., *Hierarchy Theory: The Challenge of Complex Systems*, 31–47. New York: Braziller.

Grossberg, S. 1982. *Studies of Mind and Brain*. Dordrecht: D. Reidel.

————. 1988. *Neural Networks and Natural Intelligence*. Cambridge: M.I.T. Press.

Guistino, D. de. 1975. *Conquest of Mind: Phrenology and Victorian Social Thought*. London: Croom Helm.

Hacking, I. 1983. *Representing and Intervening: Introductory Topics in the Philosophy of Natural Science*. Cambridge: Cambridge University Press.

Haeckel, E. 1866. *General Morphology of Organisms*. Berlin: Reimer.
———. 1874. *Anthropogenesis; or, the Developmental History of Man*. Leipzig: Engelmann.
Hall, T. S. 1951. *A Sourcebook in Animal Biology*. Cambridge: Harvard University Press.
———. 1969. *Ideas of Life and Matter*. 2 vols. Chicago: University of Chicago Press.
Haller, A. von. 1739. *Descriptio Foetus Bicipitis ad Pectora Connati ubi in Causas Monstrorum ex Principiis Anatomicis Inquiritur*. 2 vols. Hannover: B. Nic. Foerster.
———. 1756–1760. *Mémoires sur la Nature Sensible et Irritable des Parties du Corps Animal*. 4 vols. Lausanne: M. M. Bousquet and S. d'Arnay.
———. 1757–1766. *Elementar Physiologie Corporis Humani*. 8 Vols. Lausanne: M. M. Bousquet.
Hanson, N. R. 1958. *Patterns of Discovery*. Cambridge: Cambridge University Press.
Harden, A. 1903. Ueber alkoholische Gährung mit Hefe-Presstoff (Buchner's Zymase) bei Gegenwart von Blutserum. *Berichte der deutschen chemischen Gesellschaft* 36:715–16.
———. 1932. *Alcoholic Fermentation*. 4th ed. London: Longmans, Green, and Company.
Harden, A., and W. J. Young. 1906. The Alcoholic Fermentation of Yeast-juice, *Proceedings of the Royal Society, London* B77:405–520. Reprinted in Kalckar (1969), 20–33.
———. 1908. The Alcoholic Fermentation of Yeast-juice, Part III. *Proceedings of the Royal Society, London* B80:299–311.
Hari, R, and O. V. Lounasmaa. 1989. Recording and Interpretation of Cerebral Magnetic Fields. *Science* 244:432–36.
Harrington, A. 1987. *Medicine, Mind, and the Double Brain*. Princeton: Princeton University Press.
Hassenfratz. 1791. Mèmoire sur la Combinaison de l'Oxigéne avec le Carbone et l'Hydrogéne du sang. *Annales de Chemie et Phys* 9:261.
Hawkins, R., and E. Kandel. 1984. Is there a Cell Biological Alphabet for Simple Forms of Learning? *Psychological Review* 91:375–91.
Head, H. 1926. *Aphasia and Kindred Disorders of Speech*. 2 vols. New York: Hafner Publishing Co.
Hebb, D. O. 1949. *The Organization of Behavior*. New York: John Wiley.
Helmholtz, H. von. 1847. *Über die Erhaltung der Kraft*. Berlin: G. Reimer.
———. 1848. Ueber die Wärmeentwickelung bei Muskelaction. *Müllers Archiv für Anatomie, Physiologie, und wissenschaftliche Medizin*. In *Wissenschaftliche Abhandlungen* 2, 745–63 (Leipzig 1883).
Hermann, L. 1867. *Untersuchungen über den Stoffwechsel der Muskeln*. Berlin: Hirschwald.
Hinton, G. 1986. Learning Distributed Representations of Concepts, *Proceedings of the Eighth Annual Conference of the Cognitive Science Society*. Hillsdale, N.J.: Lawrence Erlbaum.
Holmes, F. L. 1974. *Claude Bernard and Animal Chemistry*. Cambridge: Harvard University Press.

————. 1985. *Lavoisier and the Chemistry of Life: An Exploration of Scientific Creativity*. Madison: University of Wisconsin Press.

————. 1986. Intermediary Metabolism in the Early Twentieth Century. In W. Bechtel, ed., *Integrating Scientific Disciplines*, 59–76. Dordrecht: Martinus Nijhoff.

Holyoak, K. J. 1990. Problem Solving. In D. N. Osherson and E. E. Smith, eds., *Thinking: An Invitation to Cognitive Science*, 3:117–46. Cambridge: M.I.T. Press.

Hoppe, F. 1862. Über das Verhalten des Blutfarbstoffes im Spectrum des Sonnenlichtes. *Virchows Archiv* 23:446–49.

Hoppe-Seyler, F. 1876. Ueber die Processe der Gährungen und ihre Beziehung zum Leben der Organismus. *Pflüger's Archiv* 12:1–17.

Hubel, D. H. 1979. The Brain. *Scientific American* 241:45–53.

Hull, D. L. 1974. *Philosophy of Biological Science*. Englewood Cliffs, N.J.: Prentice Hall.

————. 1976. Are Species Really Individuals? *Systematic Zoology* 25:174–91.

————. 1978. A Matter of Individuality. *Philosophy of Science* 45:335–60.

Hume, D. 1888. *A Treatise of Human Nature*. Edited by L. A. Selby-Bigge. Oxford: Clarendon Press.

Huxley, T. H. 1863. *Man's Place in Nature*. Reprinted as vol. 7 of Huxley (1896–1902).

————. 1874. On the Hypothesis that Animals are Automata, and Its History. Reprinted in Huxley (1896–1902), 1:199–250.

————. 1896–1902. *Collected Essays*. 9 vols. New York: Greenwood Press.

Hyatt, A. 1866. On the Parallelism Between the Different Stages of Life in the Individual and Those in the Entire Group of the Molluscous Order Tetrabranciata. *Memoirs of the Boston Society of Natural History* 1:193–209.

Hyman, G. T., et al. 1984. Alzheimer's Disease: Cell-Specific Pathology Isolates the Hippocampal Formation. *Science* 225:1168–70.

Issacson, R. L. 1974. *The Limbic System*. New York: Plenum Press.

Jackson, J. H. 1866. Notes on the Physiology and Pathology of Language. *Medical Times and Gazette*. Reprinted in Jackson (1958), 121–28.

————. 1873. On the Anatomical and Physiological Localisations of Movements in the Brain. In K. H. Pribram, ed., *Perception and Action*, 472–90. Baltimore: Penguin Books, 1969.

————. 1874. On the Nature of the Duality of the Brain. In Jackson (1958), 2:129–45.

————. 1882. On Some Implications of Dissolution of the Nervous System. In Jackson (1958), 2:29–44.

————. 1884. Evolution and Dissolution of the Nervous System (The Croonian Lectures), *British Medical Journal*, 1887. Reprinted in Jackson (1958), 2:45–91.

————. 1958. *Selected Writings of John Hughlings Jackson*. 2 vols. Edited by J. Taylor. New York: Basic Books.

Johnson-Laird, P. N. 1983. *Mental Models*. Cambridge: Harvard University Press.

Johnson-Laird, P. N., and P. C. Wason. 1970. A Theoretical Analysis of Insight into a Reasoning Task. *Cognitive Psychology* 1:134–48.

Kahneman, D., P. Slovic, and A. Tversky, eds. 1982. *Judgment Under Uncertainty: Heuristics and Biases*. Cambridge: Cambridge University Press.

Kalckar, H. M. 1969. *Biological Phosphorylations*. Englewood Cliffs, N.J.: Prentice Hall.

Kane, T. C., and R. C. Richardson. 1985. Regressive Evolution: An Historical Perspective. *The NSS Bulletin* 47:71–77.

Kastle, J. W. 1910. The Oxidases and Other Oxygen Catalysts Concerned in Biological Oxidations. *U.S. Hygenic Laboratory* 59:1–164.

Kauffman, S. A. 1974. The Large Scale Structure and Dynamics of Gene Control Circuits: An Ensemble Approach. *Journal of Theoretical Biology* 44:167–82.

———. 1976. Articulation of Parts Explanation in Biology and the Rational Search for Them. In M. Grene and E. Mendelsohn, eds., *Topics in the Philosophy of Biology*, 245–63. Dordrecht: D. Reidel.

———. 1984. Emergent Properties in Random Complex Automata. *Physica D* 10:145–56.

———. 1986. A Framework to Think about Evolving Genetic Regulatory Systems. In W. Bechtel, ed., *Integrating Scientific Disciplines*, 165–84. Dordrecht: Martinus Nijhoff.

———. Forthcoming. *The Origins of Order: Self-Organization and Selection in Evolution*. Oxford: Oxford University Press.

Keilin, D. 1966. *The History of Cell Respiration and Cytochrome*. Cambridge: Cambridge University Press.

Kelly, K., and C. Glymour. 1990. Getting to the Truth Through Conceptual Revolutions. In A. Fine, M. Forbes, and L. Wessels, eds., *PSA 1990: Proceedings of the 1990 Biennial Meeting of the Philosophy of Science Association*, 1:89–96. East Lansing: Philosophy of Science Association.

Kelsoe, J. R., et al. 1989. Re-evaluation of the Linkage Relationship Between Chromosome 11p Loci and the Gene for Bipolar Affective Disorder in the Old Order Amish. *Nature* 342:238–43.

Kitcher, P. 1984. 1953 and All That. A Tale of Two Sciences. *Philosophical Review* 43:335–74.

Kleist, K. 1934. *Gehirnpathologie*. Leipzig: Barth Publications.

Knoop, F. 1904. *Der Abbau aromatischer Fettsäuren in Tierkorper*. Frieburg: Kuttruff.

Kohler, R. 1971. The Background to Eduard Buchner's Discovery of Cell-free Fermentation. *Journal of the History of Biology* 4:35–61.

———. 1973a. The Background to Otto Warburg's Conception of Atmungsferment. *Journal of the History of Biology* 6:171–92.

———. 1973b. The Enzyme Theory and the Origin of Biochemistry. *Isis* 64:181–96.

———. 1974. The Background to Arthur Harden's Discovery of Cozymase. *Bulletin of the History of Medicine* 48:22–40.

———. 1982. *From Medical Chemistry to Biochemistry: The Making of a Biomedical Discipline*. Cambridge: Cambridge University Press.

Krebs, H. 1981. *Otto Warburg: Cell Physiologist, Biochemist, and Eccentric*. Oxford: Oxford University Press.

Kuhn, T. 1962. *The Structure of Scientific Revolutions*. Chicago: University of Chicago Press.

Kuhne, W. 1877. Ueber das Verhalten verschiedener organisirter und soganannte ungeformter Fermente. *Verhandlung Naturhist-Medicin. Heidelberg* 1:190–93.

Kützing, F. 1837. Microscopische Untersuchungen über die Hefe und Essigmutter, nebst mereren andern gehrigen vegetabilischen Gebilden. *Journal für praktische Chemie* 2:385–409.

Langley, P., H. A. Simon, G. L. Bradshaw, and J. M. Zytkow. 1987. *Scientific Discovery: Computational Explorations of the Creative Process.* Cambridge: M.I.T. Press.

Laudan, L. 1977. *Progress and Its Problems.* Berkeley and Los Angeles: University of California Press.

Lavoisier, A. L. 1799. *Elements of Chemistry.* 4th ed. Translated by R. Kerr. Edinburgh: Creech.

Lavoisier, A. L., and P. S. de LaPlace. 1780 [1784]. Mémoire sur la Chaleur. *Mémoires Académie Sciences*, 35–408. Reprinted in Lavoisier, *Oeuvres de Lavoisier* 2:318–33. Paris: Imprimerie Impériale, 1862.

Laymon, R. 1980. Idealization, Explanation, and Confirmation. In P. D. Asquith and R. N. Giere, eds., *PSA 1980: Proceedings of the 1980 Biennial Meeting of the Philosophy of Science Association*, 1:336–50. East Lansing: Philosophy of Science Association.

———. 1982. Scientific Realism and the Hierarchical Counterfactual Path from Data to Theory. In P. D. Asquith and T. Nickles, eds., *PSA 1982: Proceedings of the 1982 Biennial Meeting of the Philosophy of Science Association*, 2:107–21. East Lansing: Philosophy of Science Association.

———. 1989. Cartwright and the Lying Laws of Physics. *Journal of Philosophy* 86:353–72.

Lederberg, J. 1960. A View of Genetics. *Science* 131:269–76.

Leicester, H. M. 1968. *Sourcebook in Chemistry: 1900–1950.* Cambridge: Harvard University Press.

———. 1974. *The Development of Biochemical Concepts from Ancient to Modern Times.* Cambridge: Harvard University Press.

Lenneberg, E. H. 1967. *Biological Foundations of Language.* New York: John Wiley & Sons.

Lenoir, T. 1982. *The Strategy of Life.* Dordrecht: D. Reidel.

Levins, R. 1970. Complex Systems. In C. H. Waddington, ed., *Towards a Theoretical Biology*, 73–88. Edinburgh: University Press.

Lewontin, R. C. 1974. *The Genetic Basis of Evolutionary Change.* New York: Columbia University Press.

———. 1978. Adaptation. *Scientific American* 239:212–30.

Liebig, G. 1850. Über die Respiration der Muskeln. In *Mullers Archiv für Anatomie, Physiologie, und wissenschaftliche Medizin*, 393–416.

Liebig, J. 1839. Über die Erscheinung der Gährung, Faulnis und Verwesung und ihre Ursachen. *Annalen der Pharmacie* 30:250–87.

———. 1842. *Animal Chemistry or Organic Chemistry in Its Application to Physiology and Pathology.* Translated by W. Gregory. Cambridge: John Owen.

———. 1870. Über die Gährung und die Quelle der Muskelkraft. *Annalen der Pharmacie* 153:1–47, 137–228.

Lindsay, R., et al. 1980. *Application of Artificial Intelligence for Organic Chemistry: The DENDRAL Project.* New York: McGraw Hill.

Lipmann, F. 1971. *Wanderings of a Biochemist*. New York: Wiley Interscience.

Lipmann, F., and K. Lohmann. 1930. Über die Umwandlung der Harden-Youngschen Hexosediphosphorsäure und die Bildung von Kohlenhydraphosphorsäureestern in Froschmuskelextrakt. *Biochemische Zeitschrift* 222:389–403.

Lloyd, E. 1984. A Semantic Approach to the Structure of Population Genetics. *Philosophy of Science* 51:242–64.

———. 1986. Thinking About Models in Evolutionary Theory. *Philosophica* 37:87–100.

———. 1988. *The Structure and Confirmation of Evolutionary Theory*. Westport, Conn.: Greenwood Press.

———. 1989. The Semantic Approach and Its Application to Evolutionary Theory. In A. Fine and J. Leplin, eds., *PSA 1988: Proceedings of the 1988 Biennial Meeting of the Philosophy of Science Association*, 2:278–85. East Lansing: Philosophy of Science Association.

Loeb, J. 1900. *Einleitung in die vergleichende Gehirnphysiologie*. Translated by Anne L. Loeb as *Comparative Physiology of the Brain and Comparative Psychology*. New York: Arno Press, 1973.

Loftus, G. R., and E. F. Loftus. 1976. *Human Memory: The Processing of Information*. New York: John Wiley & Sons.

Lohmann, K. 1929. Über die Pyrophosphatfraktion im Muskel. *Naturwissenschaften* 17:624–25. Partially reprinted in Leicester (1968), 367–69.

———. 1931. Darstellung der Adenylpyrophosphorsäure; augeich ein Beitrag zum Chemismus der Muskelkontraktion. *Biochemische Zeitschrift* 217:162–77.

———1934/1969. Über den Chemismus der Muskel Kontraktion. *Naturwissenschaften* 22:409. Reprinted in Kalckar (1969), 57–61.

Longino, H. 1990. *Science as Social Knowledge*. Princeton: Princeton University Press.

Ludersdorff, F. W. 1846. Über die Natur der Hefe. *Ann. Physki* 67:408–11.

Ludwig, C., and A. Schmidt. 1869. Das Verhalten der Gase, welche mit dem Blut durch den reizbaren Saugethiermuskel stromen. *Ber. sachs. Ges. Akad. Wiss* 20:12–72.

Lundsgaard, E. 1932. Betydningen af Faenomenet maelkesyrefrie Muskelkontraktioner for Opfattelsen af Mulkelkontraktiones Kemi. *Danske Hospitalstidende* 75:84–95. Reprinted in Kalckar (1969), 344–53.

Luria, A. R. 1973. *The Working Brain: An Introduction to Neuropsychology*. Translated by Basil Haigh. New York: Basic Books.

Lwoff, A. 1950. *Problems of Morphogenesis in Ciliates*. New York: John Wiley and Sons.

———. 1987. *Consciousness*. Cambridge: M.I.T. Press/Bradford Books.

Lycan, W. G. 1981. Form, Function, and Feel. *Journal of Philosophy* 78:24–50.

McCauley, R. N. 1985. Concepts, Theories, and Truth. *Emory Cognition Project* (Report #8).

———. 1986. Intertheoretic Relations and the Future of Psychology. *Philosophy of Science* 48:218–27.

McClelland, J. L., and D. E. Rumelhart. 1988. *Explorations in Parallel Distrib-*

uted Processing. A Handbook of Models, Programs, and Exercises. Cambridge: M.I.T Press/Bradford Books.

McClelland, J. L., D. E. Rumelhart, and the PDP Research Group. 1986. *Parallel Distributed Processing: Explorations in the Microstructure of Cognition.* Volume 2, *Psychological and Biological Models.* Cambridge: M.I.T Press/Bradford Books.

McEvoy, J. G. 1988. The Enlightenment and the Chemical Revolution. In R. S. Woolhouse, ed., *Metaphysics and Philosophy of Science in the Seventeenth and Eighteenth Centuries,* 307–25. Dordrecht and Boston: Kluwer.

McEvoy, J. G., and J. E. McGuire. 1975. God and Nature: Priestley's Way of Rational Dissent. In R. McCormmach, ed., *Historical Studies in the Physical Sciences,* 325–404. Princeton: Princeton University Press.

Magendie, F. 1809. Quelques ideés Générales sur les Phénomènes Particuliers aux Corps Vivens. *Bulletin de Sciences de la Société Médicine d' émulation de Paris* 4. Translated in Albury (1977), 107–15.

———. 1821. Mémoire sur les Organes de l'Absorption chez las Mammitères; lu à l'Institut, le 7 août 1809, par M. Magendie, docteur en médicine. *Journal de Physiologie Expérimentale et Pathologique* 1:18–32.

———. 1822. Expériences sur les Fonctions des Racines des Nerfs qui Naissent de la Moelle épinière. *Journal de Physiologie Expérimentale et Pathologique* 4. Partially translated in Hall (1951), 305–8.

Magnus, G. 1837. Über die im blute enthaltenen Gase, Sauerstoff, Stickstoff, und Kohlensaure. *Poggendorf's Annalen der Physik und Chemie* 40:583–606.

Maienschein, J. 1978. Cell Lineage, Ancestral Reminiscence, and the Biogenetic Law. *Journal of the History of Biology* 11:129–58.

———. 1981. Shifting Assumptions in American Biology: Embryology 1890–1910. *Journal of the History of Biology* 14:89–113.

———. 1986. Preformation or New Formation—or Neither or Both. In T. J. Horder, J. A. Witkowski, and C. C. Wylie, eds., *A History of Embryology,* 73–108. Cambridge: Cambridge University Press.

———, ed. 1986b. *Defining Biology: Lectures from the 1890s.* Cambridge: Harvard University Press.

———. 1987. Heredity/Development in the United States, circa 1900. *History and Philosophy of the Life Sciences* 9:79–93.

Manasseïn, M. 1897. Zur Frage von der alcoholishen Gährung ohne lebende Hefezellen. *Berichte der deutschen chemischen Gesellschaft* 30:3061–62.

Margulis, L. 1970. *Origin of Eukaryotic Cells.* New Haven: Yale University Press.

Martin, J. R. 1989. Ideological Critiques and the Philosophy of Science. *Philosophy of Science* 56:1–22.

Matteucci, C. 1856. Recherches sur les Phenomenes Physiques et Chimiques de Contraction Musculaire. *Annales de Chimie,* 3d ser., 47:129–53.

May, R. M. 1974. Biological Populations with Non-overlapping Generations: Stable Points, Stable Cycles, and Chaos. *Science* 186:645–47.

———. 1976. Simple Mathematical Models with Very Complicated Dynamics. *Nature* 261:459–67.

Mayr, E. 1961. Cause and Effect in Biology. *Science* 134:1501–6.

Medin D. L., and M. M. Schaffer. 1978. A Context Theory of Classification Learning. *Psychological Review* 85:207–38.

Meehl, P. E. 1954. *Clinical Versus Statistical Prediction*. Minneapolis: University of Minnesota Press.

Melville, I. D. 1982. The Medical Treatment of Epilepsy: A Historical Review. In F. C. Rose and W. F. Bynum, eds., *Historical Aspects of the Neurosciences*, 127–36. New York: Raven Press.

Mendelsohn, E. 1964a. *Heat and Life*. Cambridge: Harvard University Press.

———. 1964b. The Biological Sciences in the Nineteenth Century: Some Problems and Sources. In A. C. Crombie and A. M. Haskin, eds., *History of Science* 3:39–59. Cambridge: W. Heffer.

———. 1977. Social Construction of Scientific Knowledge. In E. Mendelsohn, ed., *The Social Production of Scientific Knowledge*, 3–26. Dordrecht: D. Reidel.

Meyerhof, O. 1918. Über das Vorkommen des Coferments der alkoholischen Hefegärung im Muskelgewebe und sein mutmassliche Bedeutung im Atmungsmechanismus. *Zeitschrift für physiologische Chemie* 101:165–75.

———. 1924. *Chemical Dynamics of Life Phenomena*. Philadelphia: Lippincott.

———. 1930. *Die Chemische Vorgänge im Muskel und ihr Zusammenhang mit Arteitsleistung und Wärmebildung*. Berlin: Springer.

Meyerhof, O., and K. Lohmann. 1927. Über den Ursprung der Kontraktionswärme. *Naturwissenschaften* 15:670–71. Reprinted in Kalckar (1969), 342–43.

———. 1932. Über die Energetik der anaeroben Phosphagensynthese Kreatinphosphorsäure in Muskelextrakt. *Naturwissenschaften* 19:575–76.

Meyerhof, O., K. Lohmann, and Meyer. 1931. Über das Koferment der Milchsäure im Muskel. *Biochemische Zeitschrift* 237:437–44.

Miahle, L. 1856. *Chimie appliquée à la Physiologie et à la Thérapeutique*. Paris: Masson.

Mill, J. S. 1884. *A System of Logic*. London: Longmans.

Minsky, M., and S. Papert. 1969. *Perceptrons*. Cambridge: M.I.T Press.

———. 1974. *Artificial Intelligence*. Oregon State System of Higher Education.

Monakow, C. 1910. *Über Lokalisation der Hirnfunctionen*. Wiesbaden.

———. 1914. *Die Lokalisation in Grosshirn and der Abbau der Funktionen durch corticale Herde*. Wiesbaden.

Morgan, T. H., and C. B. Bridges. 1919. The Origin of Gynandromorphs. In Morgan and Bridges, *Contributions to the Genetics of Drosophila Melanogaster*. Washington, D.C.: Carnegie Institution Publication No. 278.

Morgan, T. H., et al. 1915. *The Mechanism of Mendelian Heredity*. London: Constable & Co.

Nachmansohn, D. 1972. Biochemistry as Part of My Life. *Annual Reviews of Biochemistry* 41:1–28.

Needham, D. M. 1937. Chemical Cycles in Muscle Contraction. In J. Needham and D. E. Green, eds., *Perspectives in Biochemistry*, 201–14. Cambridge: Cambridge University Press.

———. 1971. *Machina Carnis: The Biochemistry of Muscular Contraction in its Historical Development*. Cambridge: Cambridge University Press.

Needham, D. M., and R. K. Pillai. 1937. The Coupling of Oxido-reductions and Dismutations with Esterification of Phosphate in Muscle. *Biochemical Journal* 31:1837–51.

Neisser, U. 1967. *Cognitive Psychology*. New York: Appleton-Century-Crafts.

Neubauer, O., and K. Fromherz. 1911. Über den Abbau der Aminosäuer bei der Hefegärung. *Zeitschrift für physiologische Chemie* 70:326–50.

Neuberg, C., and J. Kerb. 1914. Über zucherfreie Hefegärungen. XII. Über Vorgänge bei der Hefegärung. *Biochemische Zeitschrift* 53:406–19.

Neuberg, C., and M. Kobel. 1925. Zur Frage der künstlichen und natürlichen Phosphoryleirung des Zuckers. *Biochemische Zeitschrift* 155:499–506.

Neuberg, C., and E. Reinfurth. 1920. Ein neues Abfangverfahren und seine anwendung auf die alkoholische Gärung. *Biochemische Zeitschrift* 106:281–91.

Newell, A. 1973. Production Systems: Models of Control Structure. In W. Chase, ed., *Visual Information Processing*. New York: Academic Press.

Newell, A., and H. A. Simon. 1961. GPS, a Program that Simulates Human Thought. In Feigenbaum and Feldman (1963), 279–96.

———. 1972. *Human Problem Solving*. Englewood Cliffs, N.J.: Prentice Hall.

Newell, A., J. C. Shaw, and H. A. Simon. 1958. Chess Playing Programs and the Problem of Complexity. *IBM Journal of Research and Development* 2:320–35. Reprinted in Feigenbaum and Feldman (1963), 39–70.

Nickles, T. 1980. Introductory Essay: Scientific Discovery and the Future of Philosophy of Science. In T. Nickles, ed., *Scientific Discovery: Logic and Rationality*, 1–60. Dordrecht: D. Reidel.

———. 1982. How Discovery Is Important to Cognitive Studies of Science. Presented at the Society for Social Studies of Science (draft).

———. 1987a. From Natural Philosophy to Metaphilosophy of Science. In R. Kargon and P. Achinstein, eds., *Kelvin's Baltimore Lectures and Modern Theoretical Physics: Historical and Philosophical Perspectives*, 507–41. Cambridge: M.I.T. Press.

———. 1987b. Methodology, Heuristics, and Rationality. In J. C. Pitt and M. Pera, eds., *Rational Change in Science*, 103–32. Dordrecht: D. Reidel.

Nisbett, R., and L. Ross. 1980. *Human Inference: Strategies and Shortcomings of Social Judgment*. Englewood Cliffs, N.J.: Prentice Hall.

Nisbett, R. E., et al. 1987. Teaching Reasoning. *Science* 238:625–31.

Oertmann, E. 1877. Über den Stoffwechsel entbluteter Frösche. *Pflügers Archiv für die gesammte Physiologie des Menschen und der Thiere* 15:381–98.

O'Keefe, J., and L. Nadel. 1978. *The Hippocampus as a Cognitive Map*. Oxford: Clarendon Press.

———. 1979. Précis of O'Keefe and Nadel's *Hippocampus as a Cognitive Map*. *Behavioral and Brain Sciences* 2:487–533.

Olby, R. 1971. "Carl F.J.E. Correns." *Dictionary of Scientific Biography* 3:421–23.

———. 1974. *The Path to the Double Helix*. Seattle: University of Washington Press.

———. 1985. *Origins of Mendelism*. 2d ed. Chicago: University of Chicago Press.

Olmstead, J.M.D. 1944. *François Magendie*. New York: Schuman.

Olton, D. S., J. T. Becker, and G. E. Handelmann. 1979. Hippocampus, Space and Memory. *Behavioral and Brain Sciences* 2:313–65.

Overton, E. 1899. Über die allgemeinen osmotischen Eigenschaften der Zelle. *Vierteljahrschrift Natur. Ges. Zurich* 44:88–135.

Pasteur, L. 1858a. Mémoire sur la Fermentation Appelée Lactique. *Annales de Chimie*, Sér. 52:404–18.

———. 1858b. Nouveaux faits Concernant l'Histoire de la Fermentation Alcoholique. *Comptes Rendus Hebdomadaires des Séances de l' Academie des Sciences, Paris* 47:1011–13.

———. 1860a. Mémoire sur la Fermentation Alcoholique. *Annales de Chimie*, 3e, Ser. 58:323–426.

———. 1860b. Note sur la Fermentation Alcoholique. *Comptes Rendus Hebdomadaires des Séances de l' Academie des Sciences, Paris* 50:1083–84.

———. 1860c. Recherches sur la Dissymétrie Moléculaire des Produits Organiques Naturels. *Societe Chimique de Paris, Leçons de Chimie*, 1–48.

———. 1872. Note sur la Mémoire de M. Liebig Relatif aux Fermentations. *Annales de Chimie et de Physique* 25:145–51.

Pattee, H. H. 1973. The Physical Basis and Origin of Hierarchical Control. In H. H. Pattee, ed., *Hierarchy Theory: The Challenge of Complex Systems*, 79–108. New York: Braziller.

Pauling, L., and M. Delbrück. 1940. The Nature of the Intermolecular Forces Operative in Biological Processes. *Science* 92:77–79.

Payen, A., and J. F. Persoz. 1833. Mémoire sur la Diastase, les Principaux Produits de ses Réactions et leurs Applciations aux arts Industriels. *Annales de Chimie et de Physique* 53:73–92.

Peters, R. A. 1957. Forty-five Years of Biochemistry. *Annual Review of Biochemistry* 26:1–16.

Pflüger, E. 1872. Über die Diffusion des Sauerstoffs, den Ort und die Gesetze der Oxydationsprocesse im thierischen Organismus. *Pflügers Archiv für die gesamte Physiologie des Menschen und der Tiere* 6:43–64.

———. 1875. Beiträge zur Lehre von der Respiration. I. Über die physiologische Verbreenung in den lebendigen Organismen. *Pflügers Archiv für die gesamte Physiologie des Menschen und der Tiere* 10:251–369.

Pinker, S., and A. Prince. 1988. On Language and Connectionism: Analysis of a Parallel Distributed Processing Model of Language Acquisition. *Cognition* 28:73–193.

Popper, K. R. 1959. *The Logic of Scientific Discovery*. New York: Harper & Row.

Popper K. R., and J. C. Eccles. 1977. *The Self and Its Brain*. New York: Springer.

Posner, M., and P. McLeod. 1982. Information Processing Models: In Search of Elementary Operations. *Annual Review of Psychology* 33:417–514.

Reichenbach, H. 1951. *The Rise of Scientific Philosophy*. Berkeley and Los Angeles: University of California Press.

Reitman, W. 1965. *Cognition and Thought*. New York: Wiley.

Rescorla, R. 1984. Comments on Three Pavlovian Paradigms. In D. L. Alkon and J. Farley, eds., *Primary Neural Substrates of Learning and Behavioral Change*. Cambridge: Cambridge University Press.

Richards, R. J. 1987. *Darwin and the Emergence of Evolutionary Theories of Mind and Behavior*. Chicago: University of Chicago Press.

Richardson, R. C. 1979. Functionalism and Reductionism. *Philosophy of Science* 46:533–58.

———. 1980. Reductionist Research Programmes in Psychology. In P. D. Asquith and R. N. Giere, eds., *PSA 1980: Proceedings of the 1980 Biennial Meeting of the Philosophy of Science Association*, 1:171–83. East Lansing: Philosophy of Science Association.

———. 1982. Grades of Organization and the Units of Selection Controversy. In P. D. Asquith and T. Nickles, eds., *PSA 1982: Proceedings of the 1982 Biennial Meeting of the Philosophy of Science Association*, 2:324–40. East Lansing: Philosophy of Science Association.

———. 1984. Biology and Ideology: The Interpenetration of Science and Values. *Philosophy of Science* 51:396–420.

———. 1986a. Language, Thought, and Communication. In W. Bechtel, ed., *Integrating Scientific Disciplines*, 263–83. Dordrecht: Martinus Nijhoff.

———. 1986b. Models and Scientific Explanations. *Philosophica* 37:59–72.

———. 1986c. Models and Scientific Idealizations. In P. Weingartner and G. Dorn, eds., *Foundations of Biology*, 109–44. Vienna: Hölder-Pichler-Tempsky.

Richardson, R. C., and T. C. Kane. 1988. Orthogenesis and Evolution in the 19th Century: The Idea of Progress in American Neo-Lamarckism. In M. Nitecki, ed., *Evolutionary Progress*, 149–67. Chicago: University of Chicago Press.

Rips, L. J. 1983. Cognitive Processes in Propositional Reasoning. *Psychological Review* 90:38–71.

Roe, S. 1981. *Matter, Life, and Generation: Eighteenth-Century Embryology and the Haller-Wolff Debate*. Cambridge: Cambridge University Press.

Rosenblat, F. 1962. *Principles of Neurodynamics*. New York: Spartan.

Sapp, J. 1987. *Beyond the Gene: Cytoplasmic Inheritance and the Struggle for Authority in Genetics*. New York: Oxford University Press.

Schoenbein, C. F. 1845. Nachträgliche Notiz über das Guajakharz. *Poggendorf's Annalen der Physik und Chemie* 67:99–100.

———. 1848. Über einige chemische Wirkungen der Kartoffel. *Poggendorf's Annalen der Physik und Chemie* 75:357–61.

Schwann, T. 1837. Vorläufige Mitteilung, betreffend Versuche über die Weingärung und Faulnis. *Poggendorf's Annalen der Physik und Chemie* 41:184–93.

———. 1839. *Mikroskopische Untersuchungen über die Übereinstimmung in der Struktur und dem Wachstum der Theire und Pflanzen*. Berlin: Sander. Translated by H. Smith as *Microscopical Researches into the Accordance in the Structure and Growth of Animals and Plants*. London: Sydenham Society, 1847.

Scriven, M. 1956. A Study of Radical Behaviorism. *Minnesota Studies in the Philosophy of Science*, 1:88–130. Minneapolis: University of Minnesota Press.

Selfridge, O. G., and U. Neisser. 1960. Pattern Recognition by Machine. *Scientific American* 203:60–68.

Selkoe, D. J., et al. 1987. Conservation of Brain Amyloid Proteins in Aged Mammals and Humans with Alzheimer's Disease. *Science* 235:873–77.

Senebier, J. 1807. *Parrorts de l'Air aven les êtres Organisés out Traités de l'Action du Poumon et de la Peau des Animaux sur l'air, comme de Celle des Plantes sur ce Fluide. Tirés des Journaux d'Observations et d'Expriences de Lazare Senebier*. 3 vols. Geneva.

Shapin, S. 1975. Phrenological Knowledge and the Social Structure of Early Nineteenth-Century Edinburgh. *Annals of Science* 32:219–43.

———. 1979. Homo Phrenologicas: Anthropological Perspectives on an Historical Problem. In B. Barnes and S. Shapin, eds., *Natural Order: Historical Studies of Scientific Culture.* London: Sage.

Simon, H. 1957. A Behavioral Model of Rational Choice. In Simon (1979), 7–19.

———. 1959. Rational Choice and the Structure of the Environment. In Simon (1979), 20–28.

———. 1966. Scientific Discovery and the Psychology of Problem Solving. In R. Colodny, ed., *Mind and Cosmos*, 22–40. Pittsburgh: University of Pittsburgh Press. Reprinted in Simon (1977), 286–303.

———. 1969. *The Sciences of the Artificial*, 2d ed. Cambridge: M.I.T. Press.

———. 1973a. The Organization of Complex Systems. In H. H. Pattee, ed., *Hierarchy Theory: The Challenge of Complex Systems*, 1–27. New York: Braziller.

———. 1973b. The Structure of Ill-Structured Problems. *Artificial Intelligence* 4:181–201.

———. 1977. *Models of Discovery*. Dordrecht: D. Reidel.

———. 1979. *Models of Thought*. New Haven: Yale University Press.

———. 1980. The Behavioral and Social Sciences. *Science* 209:72–78.

Skinner, B. F. 1938. *The Behavior of Organisms*. Englewood Cliffs, N.J.: Prentice Hall.

———. 1953. *Science and Human Behavior*. New York: Free Press.

———. 1957. *Verbal Behavior*. Englewood Cliffs, N.J.: Prentice Hall.

Slator, A. 1906. Studies in Fermentation. I. The Chemical Dynamics of Alcoholic Fermentation by Yeast. *Journal of the Chemical Society, London* 89:128–42.

Smith, C.U.M. 1982a. Evolution and the Problem of Mind: Part I, Herbert Spencer. *Journal of the History of Biology* 15:55–88.

———. 1982b. Evolution and the Problem of Mind: Part II, John Hughlings Jackson. *Journal of the History of Biology* 15:241–62.

Solomon, M. 1992. Rationality and Human Reasoning. *Philosophy of Science* 59.

Spallanzani, L. 1803. *Mémoires sur la Respiration, par Lazare Spallanzani traduits en Franais, d' apra son Manuscrit indit par Jean Senebier*. Geneva: Paschoud. English ed., London, 1804.

Spencer, H. 1851. *Social Statics*. London: Chapman.

———. *Principles of Psychology*. London: Longman.

———. *Principles of Psychology*. 2d ed. 2 vols. London: Williams & Norgate.

Sperry, R. W. 1961. Cerebral Organization and Behavior. As reprinted in Paul Tibbetts, ed., *Perception*, 81–103. New York: Quadrangle Books, 1969.

———. 1969. A Modified Concept of Consciousness. *Psychological Review* 76:532–26.

Spurzheim, Johann Gaspar. 1832. *Outlines of Phrenology*. Boston: Marsh, Capen and Lyon. As reproduced in Daniel N. Robinson, *Significant Contributions to the History of Psychology*. Washington, D.C.: University Publications of America, 1978.

Stahl, G. E. 1697/1748. *Zymotechnia Fundamentalis oder allgemeine Grunderkänntnis der Gährungs-kunst*. Frankfort and Leipzig: Montag.

Star, S. L. 1983. Simplification in Scientific Work: An Example from Neuroscience Research. *Social Studies of Science* 13:205–28.

Stokes, G. G. 1864. On the Reduction and Oxidation of the Colouring Matter of the Blood. *Proceedings of the Royal Society of London* 13:355–64.

Sturtevant, A. H. 1920. The Vermilion Gene and Gynandromorphism. *Proc. Sci. Exp. Brit. Med.* 17.

———. 1925. The Effects of Unequal Crossing Over at the Bar Locus in *Drosophila*. *Genetics* 10:117–47.

———. 1965. *A History of Genetics*. New York: Harper & Row.

Sturtevant, A. H., and G. W. Beadle. 1939. *An Introduction to Genetics*. Philadelphia and London: W. B. Saunders Co.

Tanzi, R. E., et al. 1987. Amyloid B Protein Gene: cDNA, mRNA Distribution, and Genetic Linkage Near the Alzheimer Locus. *Science* 235:880–84.

Tatum, E. L. 1959. A Case History in Biological Research. *Science* 129:1711–14.

Tatum, E. L., and G. W. Beadle. 1942. Genetic Control of Biochemical Reactions in *Neurospora:* An 'Aminobenzoicless' Mutant. *Proceedings of the National Academy of Science* 28:234–43.

Teich, M. 1981. Ferment or Enzyme: What's in a Name? *History and Philosophy of the Life Sciences* 3:193–215.

Tenon, M. M., et al. 1809. Report on a Memoir of Drs Gall and Spurzheim, Relative to the Anatomy of the Brain. *Edinburgh Medical and Surgical Journal* 5:36–66.

Thagard, P. 1980. Against Evolutionary Epistemology. In P. D. Asquith and R. N. Giere, eds., *PSA 1980: Proceedings of the 1980 Biennial Meeting of the Philosophy of Science Association*, 2:187–96. East Lansing: Philosophy of Science Association.

———. 1988. *Computational Philosophy of Science*. Cambridge: M.I.T. Press/Bradford Books.

———. 1992. *Scientific Revolutions*. Princeton: Princeton University Press.

Thénard, L. J. 1803. Mémoire sur la Fermentation Vineuse. *Annales de Chimie* 46:294–320.

———. 1818. Observations sur l'Influence de l'Eau dans la Formation des Acides Oxigénés. *Annales de Chimie et de Physique* 9:314–17.

Thompson, P. 1983. The Structure of Evolutionary Theory: A Semantic Perspective. *Studies in History and Philosophy of Science* 14:215–29.

———. 1989. The Semantic Approach and Its Application to Evolutionary Theory. In A. Fine and J. Leplin, eds., *PSA 1988: Proceedings of the 1988 Biennial Meeting of the Philosophy of Science Association*, 2:278–85. East Lansing: Philosophy of Science Association.

Thunberg, T. 1916. Über die vitale Dehydrierung der Bernsteinsäure bei Abwesenheit von Sauerstoff. *Zentralblatt für Physiologie* 31:91–93.

———. 1917. Zur Kenntnis Einwirkung tierische Gewebe auf Methylenbau. *Skandinavisches Archiv für Physiologie* 35:163.

———. 1920. Zur Kenntnis des intermediären Stoffwechsels und der dabei wirksamen Enzyme. *Skandinavisches Archiv für Physiologie* 40:9–91.

———. 1930. The Hydrogen Activating Enzymes of the Cell. *Quarterly Review of Biology* 5:318–47.

Tizard, B. 1959. Theories of Brain Localization from Flourens to Lashley. *Medical History* 3:132–45.

Tolman, E. C., B. F. Ritchie, and D. Kalish. 1946. Studies in Spatial Learning. I. Organization and the Short-Cut. *Journal of Experimental Psychology* 36:13–24.

Traube, M. 1858. Zur Theorie der Gährungen und Verwesungserscheinungen, wie der Fermentwirkungen überhaupt. *Poggendorf's Annalen der Physik und Chemie* 103:331–44.

———. 1861. Über die Beziehung der Respiration zur Muskelthätigkeit und die Bedeutung der Respiration überhaupt. *Virchows Archiv* 21:366–414.

———. 1874. Über das Verhalten der Alkoholhefe in sauerstoffgasfreien Medien. *Berichte der deutschen chemischen Gesellschaft* 7:872–87.

Traube, M. 1878. Die chemische Theorie der Fermentwirkungen und der Chemismus der Respiration. *Berichte der deutschen chemischen Gesellschaft* 11:1984–92.

Turpin, P.J.F. 1838. Mémoire sur la Cause et les Effets de la Fermentation Alcoolique et Aceteuse. *Comptes Rendus Hebdomadaires des Séances de l' Académie des Sciences, Paris* 7:369–402.

———. 1839. Über die Ursache and Wirkung der geistigen und sauren Gährung. *Justus Liebig's Annalen der Chemie* 29:93–100.

Tversky, A., and D. Kahneman. 1957. Decision under Uncertainty: Heuristics and Biases. *Science* 185:1124–31.

———. 1973. On the Psychology of Prediction. *Psychological Bulletin* 2:105–10. Reprinted in Kahneman, Slovic, and Tversky (1982), 23–31.

———. 1974. Judgment Under Uncertainty: Heuristics and Biases. *Science* 185:1124–31. Reprinted in Kahneman, Slovic, and Tversky (1982), 3–20.

Vauquelin, L. N. 1792. Observations Chimiques et Physiologiques sur la Respiration des Insectes et du Vers. *Annales de Chimie et de Physique* 12:273–91.

Warburg, O. 1910. Über die Oxydationen in lebenden Zellen nach Versuchen am Seeigelie. *Hoppe-Seyler's Zeitschrift für physiologische Chemie* 66:305–40.

———. 1911. Über Beeinflussung der Sauerstoffatmung. *Zeitschrift fur physiologische Chemie* 70:413–32.

———. 1912. Über Beziehungen zwischen Zellstuktur und biochemischen Reaktionen I. *Pflüger's Archiv für die gesammte Physiologie des Menschen und der Tiere* 145:277–82.

———. 1914a. Beitrage zur Physiologie der Zelle, insbesondere über die Oxydationsgeschwindigkeit in Zellen. *Erggebnisse der Physiologie* 14:253–337.

———. 1914b. Über die Rolle des Eisens in der Atmung des Seeigeleies nebst Bemerkungen über einige durch Eisen beschleunigte Oxydationen. *Zeitschrift für physiologische Chemie* 92:231–56.

———. 1924. Über Eisen, den sauerstoffübertragenden Bestandteil des Atmungsferments. *Biochemische Zeitschrift* 152:479–94.

———. 1925. Iron, the Oxygen Carrier of the Respiration Ferment. *Science* 61:575–82.

———. 1927. Über die Wirkung von Kohlenoxyd und Stickoxyd auf Atmung und Gärung. *Biochemische Zeitschrift* 189:354–80.

———. 1928. *Über die Katalytische Wirkung der Lebendigen Substanz.* Berlin: Springer.

————. 1929. Atmungsferment und Oxydasen. *Biochemische Zeitschrift* 214:1–3.

————. 1930. The Enzyme Problem and Biological Oxidations. *Johns Hopkins Hospital Bulletin* 46:341–58.

————. 1949. *Heavy Metal Phosphate Groups and Enzyme Action*. Oxford: Clarendon Press.

Warburg, O., and W. Christian. 1933. Über das gelbe Ferment und seine Wirkungen. *Biochemische Zeitschrift* 266:377–411.

Wason, P. C. 1966. Reasoning. In B. Foss, ed., *New Horizons in Psychology*. New York: Penguin.

Watson, J. B. 1913. Psychology as the Behaviorist Views It. *Psychological Review* 20:158–77.

————. 1924. *Behaviorism*. New York: W. W. Norton & Co.

Watson, J. D., and F.H.C. Crick. 1953. A Structure for Deolxyribose Nucleic Acid. *Nature*, 737–38.

Wernicke, C. 1874. *Der Aphasische Symptomcomplex: Eine Psychologische Studie auf Anatomischer Basis*. Breslau: Cohen & Weigert.

————. 1880. *Lehrbuch der Gehirnkrankheiten*, vol. 1. Berlin: Fischer.

————. 1886. Die neuen Arbeiten über Aphasie. *Fortschritte der Medizin*, 199–267.

————. 1906. Der Aphasische Symptomenkomplex. *Die Deutsche Klinik am Eingange die 20 Jahrhunders* 6:487. As translated in Eggert (1977).

Wieland, H. 1913. Über den Mechanismus der Oxydationsvorgänge. *Berichte der deutschen chemischen Gesellschaft* 46:3327–42. Partially reprinted in Leicester (1968), 336–40.

————. 1922. Über den Mechanismus der Oxydationsvorgänge. *Ergebnisse der Physiologie* 20:477–518.

————. 1932. *On the Mechanism of Oxidation*. New Haven: Yale University Press.

Willis, T. 1684. *Dr. Willis' Practice of Physick, etc*. Translated by S. Pordage. London: Dring, Harper, & Leigh.

Wimsatt, W. C. 1972. Complexity and Organization. In K. Schaffner and R. S. Cohen, eds., *PSA 1972: Proceedings of the 1972 Biennial Meeting of the Philosophy of Science Association*, 2:67–86. Dordrecht: D. Reidel.

————. 1976. Reductionism, Levels of Organization, and the Mind-Body Problem. In G. Globus, G. Maxwell, and I. Savodnik, eds., *Brain and Consciousness: Scientific and Philosophic Strategies*, 199–267. New York: Plenum Press.

————. 1980a. Reductionistic Research Strategies and Their Biases in the Units of Selection Controversy. In T. Nickles, ed., *Scientific Discovery: Case Studies*, 213–59. Dordrecht: D. Reidel.

————. 1980b. Randomness and Perceived-Randomness in Evolutionary Biology. *Synthese* 43:287–329.

————. 1981. Units of Selection and the Structure of the Multi-Unit Genome. In P. D. Asquith and R. N. Giere, eds., *PSA 1980: Proceedings of the 1980 Biennial Meeting of the Philosophy of Science Association*, 2:122–83. East Lansing: Philosophy of Science Association.

————. 1986a. Forms of Aggregativity. In A. Donagan, A. N. Perovich, and M. V. Wedin, eds., *Human Nature and Natural Knowledge*, 259–91. Dordrecht: D. Reidel.

Wimsatt, W. C. 1986b. Developmental Constraints, Generative Entrenchment, and the Innate-Acquired Distinction. In W. Bechtel, ed., *Integrating Scientific Disciplines*, 185–208. Dordrecht: Martinus Nijhoff.

Wolkow, M., and E. Baumann. 1891. Über das Wesen der Alkaptonurie. *Hoppe-Seyler's Zeitschrift für physiologische Chemie* 15:228–85.

Wollaston, W. H. 1804. On Certain Chemical Effects of Light. [A letter to Nicholson.] *Journal of Natural Philosophy* 6:293–97.

Wright, Sewall. 1934. Physiological and Evolutionary Theories of Dominance. *American Naturalist* 68:24–53.

———. 1941. The Physiology of the Gene. *Physiological Reviews* 21:487–523.

Young, R. M. 1970. *Mind, Brain and Adaptation in the Nineteenth Century*. Oxford: Clarendon Press.

Young, W. J. 1907. The Organic Phosphorus Compound Formed by Yeast-juice from Soluble Phosphates. *Proceedings of the Chemical Society* 23:65–66.

Index

Abrahamsen, A. A., 255
aggregativity, 25–26, 71–72, 92, 199
Albury, W. R., 100, 103, 250
alkaptonuria, 182–85
Alkon, D. L., 127
Allen, G. E., 178, 179, 253
Alzheimer's disease, 120–21
analogy, 17–18
analytic method, 18–20, 21, 138, 151–52, 171–72
Anderson, J. R., 211
Angell, J. R., 42, 246
aphasia, 128–33, 139–40, 203–4, 208–9
Appert, N., 105
arationality assumption, 9
associationism, 42–43, 131–32, 139, 204
Atmungsferment, 73, 82–84
ATP/ADP Cycle, 166–70
autonomy, 175–77, 179–81

Bach, A., 76, 85
BACON, 13–15, 121, 236
Barnes, B., 120
Barnes, D. M., 66
Bateson, W., 177, 182, 253
Battelli, F., 77
Baumann, E., 182–83
Beadle, G. W., 123–24, 179, 181–82, 185–92, 239, 243
Beatty, J., 232
Bechtel, W., 21, 26, 51, 142, 151, 168, 229, 237, 240, 252, 255
Becker, J. T., 134, 252
behaviorism, 42–44
Bell, C., 139
Bernard, C., 21, 54, 55–56, 60–61, 114, 155–56, 239, 247–48, 250
Berthelot, M., 53, 100, 112–13, 153
Berzelius, J. J., 106, 109, 246
biases of reasoning, 5
Bichat, X., 38, 99–104
biochemical genetics, 123–24, 174, 188
Boakes, R., 246
Bogen, J., 237–38, 242
Bouillaud, J. B., 128, 203
Bower, G., 251

Bowler, P., 50
Braine, M.D.S., 13
Brandon, R., 46, 246
Brentano, F., 42
Bridges, C. B., 175–76, 178
Broca, P., 127–30, 133, 138, 199, 203, 205–6, 209, 231, 252
Buchanan, G. G., 13–15
Buchner, E., 154–55, 158–60, 161, 163, 168
Buchner, H., 154–55
Burian, R., 253, 255

Cagniard-Latour, C., 105, 209
Carlson, N., 180
Cartesianism, 66–67, 95
Cartwright, N., 231, 238
Charness, N., 245
Chase, W. G., 28
Cheng, P. W., 13
Chi, M.T.H., 13
Chomsky, N., 15, 48–49, 126, 211, 247
cognitive maps, 122, 135
Colin, J. J., 250
Combe, G., 67, 69
complex systems, 18, 27. See also integrated systems
complexity, 21–23
component systems, 26
composite systems, 26
confirmation, 4–5, 245
Connant, J. B., 109
connectionism, 139, 199–229
constraints on theories, 5, 8, 156–58, 173–74, 234–43; bottom-up, 127, 138–46; multiple, 145; operational, 158–60, 183–84, 187–89, 234, 239–41; phenomenological, 234, 237–39; physical, 160–62, 184, 187, 189, 234, 242–43; psychological 234–36; top-down, 127–38, 210
control: internal, 47–51; external, 41–47; locus of, 34–36, 39–41, 59–64, 233
Cooter, R., 65–66, 248
Cope, E. D., 50–51, 247
correlational methods, 38, 64, 90, 122, 138–40
Correns, C.F.J.E., 177

Cranefield, P. F., 148
craniology, 66, 70
Crick, F.H.C., 29
Cuénot, L., 181, 254
Culotta, C. A., 55, 247
Cummins, R., 18
Cuvier, G., 33, 66, 95, 115–16

Dakin, H. D., 77
Darden, L., 4, 154, 231–32, 245, 253
Darwin, C., 44–46, 50, 149, 206, 247
Darwinian theory, 37
Davies, J. D., 66, 248
Davy, H., 74
Davy, J., 54
Dawes, R. M., 245
decomposability, 24, 39, 64, 119, 145–48;
    evidence for, 131, 137–38; and linearity,
    227–34; near, 26, 149; simple, 25, 26, 150
decomposition, 7, 18–19, 23, 28, 31–32, 64,
    88, 119, 130–33, 235–36
deficit studies. See experimental techniques
Delbrück, M., 29
DENDRAL, 13–15, 236
Dennett, D. C., 18
Descartes, R., 93–94, 97, 98, 248. See also
    Cartesianism
Desmazieres, J. B., 250
Deuticke, H. J., 164
developmental genetics, 181–88
Dewey, J., 42
Dijksterhuis, E. J., 197
discovery, 3–10; AI programs, 13–16;
    choice-points in, 35, 59, 89, 115–16, 146–
    47, 171–72, 233–34; as heuristic search,
    15; and justification, 3–4, 6; and problem
    solving, 21; psychological constraints on,
    5–6, 7, 9, 235–39
dissolution (loss of higher function), 209
Dixon, M., 249
Donaldson, H. H., 246
Drosophila, 175, 178–81, 185–88, 192
du Bois-Reymond, E., 155
dualism, 94
Duclaux, E., 159
Duhem, P., 238
Dutrochet, R.J.H., 54

East, E. M., 33
Eccles, J. C., 94
Einbeck, H., 77
Egeland, J. A., 120

Eggleton, P., 166
Eggleton, G. P., 166
Einbeck, H., 77
Ellis, D., 54
Elliott, K.A.C., 249
Embden, G., 163–72
emergence, 94, 200, 229
empirical constraints, 4, 5
Engel, H., 163
Engler, C., 85
enzymes, 72, 124, 152, 184, 186–92, 249;
    dehydrogenases, 74–80; oxygen-activat-
    ing, 80–87
Ephrussi, B., 123–24, 181, 185–88
epigenesis, 49–50, 93
epilepsy, 204–8
equipotentiality, 96, 137
excitatory studies. See experimental tech-
    niques
experimental techniques: deficit studies 19,
    61, 65, 80, 84–86, 131, 157, 166, 240–41;
    excitatory studies, 20, 65, 80, 91, 241;
    transplant studies, 185–86
expertise, 13, 28
explanation, 6–7, 12, 22, 38, 40; mechanis-
    tic, 17–23, 38, 231; spurious, 21
external control. See control

Farley, J., 127
Farr, M. J., 13
Faust, D., 5, 245
Feigenbaum, E. A., 13–15, 245
fermentation, 104–13, 152–72
Fernbach, A., 159
Ferrier, D., 251
first-order independence, 146
Fisher, R. A., 46–47
Fiske, C. H., 166
Fletcher, W. M., 156, 164
Florkin, M., 273
Flourens, J.P.M., 38, 66, 95–99, 114, 119,
    145, 199, 204, 240, 250, 254
Fodor, J. A., 126, 210, 222, 239
Fowler, O. S., 67, 69
Fraassen, B. van, 232, 238
Fritsch, G., 241
Fruton, J., 105
functional analysis, 89–90

Galison, P., 244
Gall, F. J., 38, 66–72, 89–91, 95, 97–99, 119,
    128, 138, 145, 230–31, 248, 250, 251

Garrod, A., 181–85, 242, 253–54
Gay-Lussac, J. L., 100, 105
Gayon, J., 253
genetic regulation, 200, 223–27
Geschwind, N., 127
Ghiselin, M. 59
Giere, R., 4, 232
Gilmartin, K., 28
Glas, E., 59
Glaser, H. R., 13
Gluck, M., 251
glycolysis, 152–72
Glymour, C., 8
Goldgaber, D., 121
Goldman, A. I., 10
Gould, S. J., 47, 50, 247, 254
Graham, T., 54
Gregory, R. L., 17, 19
Grene, M., 3
Griesbach, E., 163
Grobstein, C., 30–31
Guistino, D. de, 66, 248

Hacking, I., 230, 238
Haeckel, E., 36, 50, 247
Hahn, M., 155
Haldane, J.B.S., 181
Haller, A. von, 101–2
Hamilton, W., 65
Handelmann, G. E., 134, 252
Hanson, N. R., 1, 237
Harden, A., 159, 161–63, 253
Harrington, A., 19, 248
Harvey, W., 93
Hawkins, R., 127
Head, H., 255
Hebb, D. O., 213
Helmholtz, H. von, 55, 248
Hermann, L., 155
heuristics, 8–16, 23, 228, 230–31, 235–36;
    domain specific, 11; failure of, 9–10, 37; in
    language learning, 15
hierarchical systems, 27–31, 150–51, 203–9
Hill, A. V., 164
Hinton, G., 218–22
hippocampus, 140–44
history of science, 8–10
Hitzig, E., 241
holism, 94–95
Holmes, F. L., 100, 247, 249
Holyoak, K. J., 11, 13
Hopkins, F. G., 156, 164

Hoppe-Seyler, F., 73, 153
Hubel, D., 117
Hull, D. L., 59, 242
human reasoning, 5–7, 10, 16
Hume, D., 42, 205
Huxley, T. H., 94
Hyatt, A., 50–51, 247
Hyman, G. T., 121

independence, 39
independent assortment, 174, 177–78
inhibitory studies. See experimental tech-
    niques, deficit studies
integrated systems, 18, 26, 149–72, 202–29
interaction, 27–28, 32, 63, 171–72
interfield theories, 231–32
internal control. See control
Isaacson, R. L., 137

Jackson, J. H., 199–200, 203–10, 231, 241,
    251, 254–55
James, W., 42
Johnson-Laird, P. N., 5, 7
justification, 4, 5

Kahneman, D., 5, 8–9, 245
Kalberlah, F., 163
Kalish, D., 134–35
Kandel, E., 127
Kastle, J. W., 73
Kauffman, S. A., 200–201, 223–27, 246, 255
Keilin, D., 247, 249
Kelly, K., 8
Kelsoe, J. R., 251
Kerb, J., 150, 242
Kimble, D. P., 136–37
Kitcher, P., 242–44, 253
Knoop, F., 77, 158
Kohler, R., 80–81, 154–55
Kraft, G., 164
Krebs, H., 249
Kuhn, T., 237
Kuhne, W., 153
Kützing, F., 105

LaGrange, J. L. 52–53
Langley, P., 14
language learning, 48–49
LaPlace, P., 52, 240
Laudan, L., 9
Lavoisier, A. L., 38, 52, 53, 100, 104–5, 240
laws, scientific, 232, 256

Laymon, R., 7
Lederberg, J., 193–94
Lenneberg, E. H., 137, 251
Lenoir, T., 95
Levins, R., 25, 26, 31
Lewontin, R. C., 21–23, 47
Liebig, G., 55
Liebig, J., 20–21, 55, 94, 106, 109–12, 153, 239, 246–48, 252
Lindsay, R., 13
linearity, 18, 23, 150–51, 157–58, 162, 188–89; compromise of, 162–63, 165–68
Lipmann, F., 164
Lloyd, E., 232
locale systems, 134
localization, 7, 18, 19, 24, 119; complex or indirect, 24, 122–48, 192–93; direct or simple, 24, 36–37, 63–65, 73, 81, 87–100, 119, 152, 154, 192; false or vacuous, 32, 142; of function, 59–62, 90, 126–28, 131, 206–7, 209–10; insufficiently constrained, 119; minimally constrained, 63, 138; multiply constrained, 121–24, 125–38, 144–46, 151–62
Locke, J., 205
locus of control. See control
Loeb, J., 80
Lohmann, K., 164, 166–67
Longino, H., 9, 245
Ludersdorff, F. W., 153
Ludwig, C., 55, 57, 60
Lundsgaard, E., 166
Lwoff, A., 197
Lycan, W. G., 18, 210

McCauley, R. N., 7, 127
McClelland, J. L., 255
McClintock, B., 243
McEvoy, J. G., 247
McGuire, J. E., 247
MacLeod, P., 18
Magendie, F., 54, 66, 103–4, 139, 250
Magnus, G., 54
Maienschein, J., 50
Manasseïn, M., 153
manic depression, 120
Margulis, L., 26
Martin, J. R., 9
Maull, N., 231–32, 253
May, R. M., 16
Mayr, E., 46
mechanism, 17–32, 37–38, 66, 121–24, 198,

222–23; beyond classical, 227–29; causal models, 231–33; opposition to, 93–116. See also explanation
Meehl, P. E., 245
Meisenheimer, J., 155, 159
Melville, I. D., 204
memory, 134–38, 140–44
Mendelian genetics, 123–24, 174–81, 188
Mendelsohn, E., 66
mentalism, 47
Meyer, R., 167
Meyerhof, O., 82, 163–72
Meynert, T., 138, 145
Mill, J. S., 231
Minsky, M., 13, 218
misplace system, 144
models, 232
model systems, 20–21, 74–76, 82–83, 85
modularity, 24–25, 126
Morgan, T. H., 175–76, 178–81, 192, 243, 254
Muller, H. J., 103, 178, 243, 254

Nachmansohn, D., 165
Nadel, L., 122, 134–38, 140–46, 151, 240–41, 252
nativism, 15, 49
Needham, D. M., 165, 167, 253
Neisser, U., 211
neo-Lamarckism, 50–51
Neubauer, O., 159
Neuberg, C., 26, 159–62, 242, 253
Neurospora, 188–92
Newell, A., 11–13, 211
Nickles, T., 4, 14
Nisbett, R., 5
nonlinear organization, 202
normative considerations, 10

Oertmann, E., 58
O'Keefe, J., 122, 134–38, 140–46, 151, 240–41, 252
Olby, R., 177, 181, 182, 186, 253
Olmstead, J.M.D., 104
Olton, D. S., 134, 251–52
operational constraints. See constraints
organization: facultative, 30–31; forms of, 18–23; horizontal 27–28; interaction, 25, 199; obligate, 30–31
organology, 67, 70–71
orthogenesis, 36, 206
oxidation, 52, 72–84. See also respiration

Papert, S., 13, 218
Pasteur, L., 38, 99, 104–13, 119, 153, 240, 252
Pattee, H. H., 150, 252
Pauling, L., 29
PDP (parallel distributed processing systems), 210–23
Peters, R. A., 170
Pflüger, E., 57–59, 60, 61, 72f
phenomenological constraints. *See* constraints
phrenology, 65–72, 89; Flourens' opposition to, 95–99, 248
physical constraints. *See* constraints
Pillai, K., 167
Pinker, S., 222
place system, 143
Popper, K. R., 3, 4, 94
position effect, 178
Positivism, 3–4, 6
Posner, M., 18
preformationism, 49–50, 93
Prince, A., 222
production systems, 211–12
psychological constraints. *See* constraints
psychologism, 7
Pylyshyn, Z., 222, 239

rationality, 10–16
reduction, 242
Reichenbach, H., 3, 245
Reinfurth, E., 253
Reitman, W., 15
respiration: controversy over locus of, 51–59; intracellular localization, 72
Richards, R. J., 254
Richardson, R. C., 7, 9, 27, 30, 206, 229, 247, 251, 255
Rips, L. J., 13
Ritchie, B. F., 134–35
Roe, S., 250
Rosenblat, F., 211, 218
Ross, L., 5
Rumelhart, D. E., 255

Sapp, J., 245, 253
Saunders, E. R., 182
Schmidt, A., 57
Schmitz, E., 163
Schoen, M., 159
Schoenbein, C. F., 72–73
Schwann, T., 38, 51–52, 99, 105–9, 153
Scriven, M., 48

Sedgwick, A., 65, 66
self-organizing systems, 22, 150
self-regulating systems, 22, 102
Selfridge, O. G., 211
Selkoe, D. J., 121
Senebier, J., 53
semantic view of theories, 232
Shapin, S., 66
Shaw, J. C., 12
Simon, H., 1, 4, 11–13, 15, 18, 25–32, 28–31, 117, 149, 229
Skinner, B. F., 43–44, 48
Slator, A., 159
Smith, C.U.M., 254
Solomon, M., 8–9
Spallanzani, L., 53
Spencer, H., 42, 71, 205–6, 254
Sperry, R. W., 94
Spurzheim, J. G., 66–72, 95, 98, 115–16, 230–31, 248
Stahl, G. E., 250
Star, S. L., 254
Stern, L., 77
Stokes, G. G., 247
Sturtevant, A. H., 178–81, 243
Subbarow, Y., 166
Sutherland, G. L., 13–15
synthetic method, 18, 20–21, 152, 171–72

Tanzi, R. E., 121
Tatum, E. L., 123–24, 187–89, 239
taxon system, 134
Teich, M., 153
Tempus and Hora, 30, 252
Thagard, P., 4, 66
Thénard, L. J., 105
Thompson, P., 232
Thunberg, T., 26, 73, 77–80, 85, 158
Titchener, E. B., 42
Tolman, E. C., 134–35
transplant studies. *See* experimental techniques
Traube, M., 73, 112–13, 153
Turpin, P.J.F., 106
Tversky, A., 5, 8–9, 245–46

Valentine, E., xi
values, 9
Vauquelin, L. N., 53
vitalism, 94–95, 111–15; in explaining fermentation, 104–13; in opposition to mechanistic physiology, 99–104

Warburg, O., 73, 80–91, 148, 240, 249
Wason, P. C., 5
Watson, J. B., 42–44, 246–47
Watson, J. D., 29
Wernicke, C., 122, 127, 130–33, 139–40, 145, 231, 251, 254
Whewell, W., 66
Wieland, H., 73–77, 79–80, 85–91, 148, 249
Willis, T., 250
Wimsatt, W. C., 6, 14, 25, 27, 31, 39–40, 127, 242, 246

Wohler, F., 106, 111
Wolkow, M., 182–83
Woodward, J., 237–38, 242
Wright, S., 177, 181, 192–93
Wundt, W., 42

Young, R. M., 65–66, 68, 250, 254
Young, W. J., 161–63

Zallen, D., 253
Ziskin, J., 245
zymase, 168